Nanostructure Science and Technology

A volume in the Nanostructure Science and Technology series. Further titles in the series can be found at:
http://www.springer.com/series/6331

Thomas J. Webster

Editor

Safety of Nanoparticles

From Manufacturing to Medical Applications

 Springer

Editor
Thomas J. Webster
Brown University
Providence, RI
USA
Thomas_Webster@brown.edu

Series Editor
David J. Lockwood
National Research Council
Ottawa, Ontario
CANADA

ISSN: 1571-5744
ISBN: 978-0-387-78607-0   e-ISBN: 978-0-387-78608-7
DOI 10.1007/978-0-387-78608-7

Library of Congress Control Number: 2008937526

This book is dedicated to the next generation of learners,
particularly my daughters Mia, Zoe, and Ava.

Whatever the challenge, we will find solutions.

Contents

Contributors

Kimberly W. Anderson
Department of Chemical and Materials Engineering, University of Kentucky, Lexington, KY 40506–0046, USA

Liming Dai
Department of Chemical and Materials Engineering, School of Engineering and UDRI, University of Dayton, 300 College Park, Dayton, OH 45469-0240, USA

Lisa DeLouise
Departments of Dermatology, Biomedical Engineering and Environmental Medicine, University of Rochester Medical Center, Rochester, NY, USA

Alison Elder
Departments of Dermatology, Biomedical Engineering and Environmental Medicine, University of Rochester Medical Center, Rochester, NY, USA

Omid C. Farokhzad
Laboratory of Nanomedicine and Biomaterials, Department of Anesthesia, Brigham and Women's Hospital, Harvard Medical School, Boston, MA, USA; MIT-Harvard Center for Cancer Nanotechnology Excellence, Massachusetts Institute of Technology, MA, USA

Frank Gu
Harvard-MIT Division of Health Sciences and Technology, Massachusetts Institute of Technology, MA, USA; MIT-Harvard Center for Cancer Nanotechnology Excellence, Massachusetts Institute of Technology, MA, USA

Anand Gupta
Nanocopoeia, Inc. 1246 W. University Avenue, Suite 463, St. Paul, MN 55104, 651-209-1184, 651-209-1187, USA, anand.gupta@nanocopoeia.com

J. Zach Hilt
Department of Chemical and Materials Engineering, University of Kentucky, Lexington, KY 40506-0046, USA, hilt@engr.uky.edu

Robert A. Hoerr

Nanocopoeia, Inc. 1246 W. University Avenue, Suite 463, St. Paul, MN 55104, 651-209-1184, 651-209-1187, USA, bob.hoerr@nanocopoeia.com

Yiling Hong

Department of Biology, University of Dayton, Dayton, OH 45469, USA

Kang Moo Huh

School of Applied Chemistry and Biological Engineering, Chungnam National, University, Daejeon 305-764, Korea

Saber M. Hussain

Applied Biotechnology Branch, Human Effectiveness Directorate, Air Force Research Laboratory, Wright-Patterson AFB, OH 45433-5707, USA

Tatsuhiro Ishida

Department of Pharmacokinetics and Biopharmaceutics, Subdivision of Biopharmaceutical Sciences, Institute of Health Biosciences, The University of Tokushima, 1-78-1, Sho-machi, Tokushima 770-8505, Japan

Yuhui Jin

Department of Chemistry, University of North Dakota, Grand Forks, ND 58202, USA, yuhui.jin@und.nodak.edu

Jay Johnson

Department of Chemical and Materials Engineering, School of Engineering and UDRI, University of Dayton, 300 College Park, Dayton, OH 45469-0240, USA

Sungwon Kim

Department of Industrial and Physical Pharmacy, Purdue University, West Lafayette, IN 47906, USA

Hiroshi Kiwada

Department of Pharmacokinetics and Biopharmaceutics, Subdivision of Biopharmaceutical Sciences, Institute of Health Biosciences, The University of Tokushima, 1-78-1, Sho-machi, Tokushima 770-8505, Japan, hkiwada@ph.tokushima-u.ac.jp

Jin-Kyu Lee

Materials Chemistry Laboratory (MCL), School of Chemistry & Molecular Engineering, Nano Systems Institute-National Core Research Center, Seoul National University, Sillim-dong, Kwanak-gu, Seoul 151-747, Korea

Yong-kyu Lee

Department of Chemical and Biological Engineering, Chungju National University, Chungbuk 380-702, Korea

Michael J. Matuszewski
Nanocopoeia, Inc. 1246 W. University Avenue, Suite 463, St. Paul, MN 55104,
651-209-1184, 651-209-1187, USA, mike.matuszewski@nanocopoeia.com

Samantha A. Meenach
Department of Chemical and Materials Engineering, University of Kentucky,
Lexington, KY 40506-0046, USA

Luke Mortensen
Departments of Dermatology, Biomedical Engineering and Environmental
Medicine, University of Rochester Medical Center, Rochester, NY, USA

Eiji Ōsawa
NanoCarbon Research Institute, Ltd., Kashiwa-no-ha, Chiba 277-0882, Japan

Rajesh A. Pareta
Division of Engineering, Brown University, 182 Hope Street, Providence, RI 02912,
USA, Rajesh_Pareta@brown.edu

Kinam Park
Department of Industrial and Physical Pharmacy, Purdue University, West
Lafayette, IN 47906, USA; Department of Biomedical Engineering, Purdue
University, BMED Building, 206 S. Intramural Drive, West Lafayette, IN 47907,
USA, kpark@purdue.edu

G. L. Prasad
Associate Professor, Department of Physiology, Temple University School
of Medicine, 3400 North Broad Street, OMS228, Philadelphia, PA 19140, USA,
glprasad@temple.edu

Amanda M. Schrand
Department of Chemical and Materials Engineering, School of Engineering
and UDRI, University of Dayton, 300 College Park, Dayton, OH 45469-0240, USA

John J. Schlager
Applied Biotechnology Branch, Human Effectiveness Directorate, Air Force
Research Laboratory, Wright-Patterson AFB, OH 45433-5707, USA

Andrew Zhuang Wang
Laboratory of Nanomedicine and Biomaterials, Department of Anesthesia,
Brigham and Women's Hospital, Harvard Medical School, Boston, MA, USA;
Radiation Oncology Program, Harvard Medical School, Boston, MA, USA;
Harvard-MIT Division of Health Sciences and Technology, Massachusetts Institute
of Technology, MA, USA; MIT-Harvard Center for Cancer Nanotechnology
Excellence, Massachusetts Institute of Technology, MA, USA

Xiaohong Wei
College of Pharmaceutical Science, Zhejiang University, 310058, China;
Department of Industrial and Physical Pharmacy, Purdue University, West
Lafayette, IN 47906, USA

Xiaojun Zhao
Department of Chemistry, University of North Dakota, Grand Forks, ND 58202, USA, jzhao@chem.und.edu

Lin Zhu
Department of Biology, University of Dayton, Dayton, OH 45469, USA

Chapter 1
Developing Practices for Safe Handling of Nanoparticles and Nanomaterials in a Development-Stage Enterprise: A Practical Guide for Research and Development Organizations

Robert A. Hoerr, Anand Gupta, and Michael J. Matuszewski

Abstract There is a need for well thought-out procedures to be developed for synthesizing nanomaterials for associated industries. In academia, where the quantity involved for these nanomaterials is usually small, with limited exposure potential, relative risk is likely unquantifiable. We have generated various protocols with different developmental phases in our industry. These steps involve aerosol monitoring, efficient filters, and enclosures. These guidelines could be utilized by numerous start-ups and scaling-up nanomaterials associated industries.

Contents

R.A. Hoerr (✉)
Nanocopoeia, Inc., 1246 W. University Avenue, Suite 463, St. Paul, MN 55104
e-mail: bob.hoerr@nanocopoeia.com

T.J. Webster (ed.), *Safety of Nanoparticles*, Nanostructure Science and Technology,
DOI 10.1007/978-0-387-78608-7_1, © Springer Science+Business Media, LLC 2009

1.1 Introduction

With the explosion in nanotechnology research across many fields, a large number of early stage ventures are participating in the development of novel, nanoscale materials. At the earliest stages of development, very little may be known about the properties of materials that are being used or generated during research. Personnel involved may be graduate students or technicians with limited training in safe laboratory practices. Quantities of materials tend to be small at this stage, with limited exposure potential, but as development proceeds, scale-up of production follows and creates new levels and types of exposure. Resources for evaluating safety of novel materials are typically limited in development stage settings; thus, while exposure levels may be low, the relative risk of the materials is likely unquantifiable. This argues for well thought-out procedures to be developed in advance of working with nanoscale materials and regular monitoring of the work environment as a development program proceeds. This chapter summarizes the authors' experiences in a start-up company that is developing a university-licensed nanoparticle production technology and provides more general guidance that may be useful in developing safe handling practices for working with these novel materials in academic laboratories as well as in early stage companies working with these novel materials.

Nanoparticles are defined as small-scale substances (<100 nm) with unique properties while nanomaterials are defined as a diverse class of small-scale (<100 nm) substances formed by molecular-level engineering to achieve unique mechanical, optical, electrical, and magnetic properties (Tsuji et al. 2006).

1.2 ElectroNanospray™ Generated Nanomaterials

The ElectroNanospray™ process for generating nanoparticles was invented at the University of Minnesota by Doctors Chen and Pui (Chen and Pui 1997, 2000). By passing solutions or suspensions of compounds through a novel co-axial capillary spray nozzle and exposing the nozzle to high voltages (typically 3–10 kV), a high velocity spray stream of highly charged nanoparticles is generated. The nanoparticles may be simple or complex, including core-shell structures, and they may be captured on a surface to create a nanostructured film. Particle size is controlled by flow rate, voltage, and conductivity of the sprayed solution, but the size may range from 10–300 nm. The particles are generated and form an aerosol stream, but because they are charged and flow within the electric field, they tend to be captured as they travel to a grounded substrate or an alternate charged surface. Particles can also be neutralized by ionizing sources and maintained in the aerosol stream.

The technology has broad applications and is being used for a variety of purposes and with a wide range of materials. Examples include:

- Coating coronary stents with drug and drug-eluting polymer
- Nanoformulating poorly water-soluble drugs with excipients to promote better water solubility or suspendability
- Transfecting cultured cells with DNA
- Aerosolizing nanoparticles for pulmonary delivery in animal nanotoxicology studies.

As a result, the types of materials being used with the ElectroNanospray™ process are quite diverse. They range from small organic molecules (pharmaceutical agents like the steroid dexamethasone, the calcium channel blocker nifedipine, and the anticonvulsant carbamazapine) to biomolecules (peptides, proteins, and DNA), as well as more exotic materials like carbon nanofibers and quantum dots.

The possibilities for workplace exposure come at multiple points: from the time raw materials and chemicals are sourced from a vendor and taken into inventory, when first material solutions are prepared, during spray operations, and during material characterization that is accomplished either by imaging or chemical analysis. For example, the use of High Performance Liquid Chromatography (HPLC) has been shown to provide excellent data on the concentration of <200 nm particles obtained during the ElectroNanospray™ process. The sites of exposure include storage containers, fume hoods, benchtops, spray equipment, and samples. The routes of exposure are topical, via the skin, eyes or mucous membranes, inhalation, via the lung, and oral, via inadvertent ingestion or contact of hands or clothing with the mouth. While this seems very logical, designing a workplace setting where these are minimized is a challenge, in part, because the amounts of material in the immediate work environment are very small and difficult to detect in routine monitoring studies.

For purposes of process development and validation, we have worked whenever possible with model compounds for which a significant amount of information related to safety and potential toxicology is known. For example, while developing model drug-eluting coatings for coronary stents based on a new polymer system, we selected drugs that were well characterized and for which significant information existed in the published scientific literature. We began with a steroid, dexamethasone, and only after understanding how this behaved in our system did we move to working with more potent drugs like paclitaxel, which is known to affect cell proliferation and is commonly used as a breast cancer treatment. Both drugs are approved pharmaceutical agents and their pharmacology and toxicology is well understood. Routine assays exist for their analysis, meaning that a method is available for quantifying the amount of material found at various points throughout the spray work cycle. We limit exposure by making sure that the amounts of material that we work with at any one time are well below the level of a single human dose.

At times, this is not possible because the compound of interest is a new molecule with significant clinical use potential but no toxicology data. In this case, the need for precaution is the highest because there is insufficient information to gauge the risk of exposure. This is akin to the use of universal precautions against HIV exposure, where every individual and biological material is assumed to be infected and exposure is minimized at each encounter by appropriate protective equipment and clothing as well as proper handling and disposal of syringes, needles, and other material that has been in contact with the individual or material.

1.2.1 Decision Algorithm for Working with New Materials

When introducing a new material, especially a nanomaterial, into the laboratory, a few things must be considered. First, what are the properties of the new material? This would include physical, chemical, and toxicological properties. If this information is known, how should it be handled? Unless a standard operating procedure is already in place for handling these new materials, the material safety data sheet is a good place to obtain pertinent information for proper handling. If the properties of the material are unknown, then it would be advisable to take every available precaution until enough information has been gathered to produce an acceptable operating procedure. The decision algorithm we have developed for handling nanomaterials is shown in Fig. 1.1.

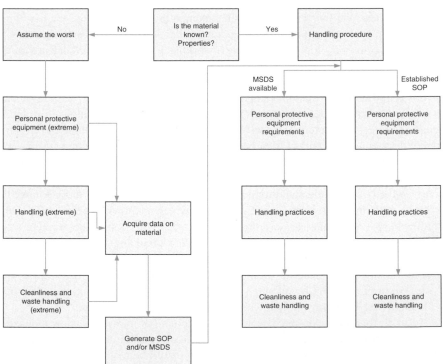

Fig. 1.1 Guidelines for safe handling of nanomaterials in the industrial research and development setting

1.3 Evolving Safe Handling Practices

As our R&D program has developed, we have instituted a variety of modifications in our ElectroNanospray nanoparticle production system. The original device was very simple with an exposed spray nozzle directed toward a flat target. It has evolved to a sophisticated system with multiple controls and containment features. Many of the design aspects have been driven by our desire to ensure safe material handling and containment of an airborne stream of nanoparticles. Table 1.1 summarizes the phases of design evolution and the related safety issues that were addressed during each phase.

Table 1.1 A case study of the evolution of safety features designed in a nanoparticles production system: from the research laboratory to a commercial production facility

Phase of design evolution	System schematic
Phase I: Open capillary spray system with a high voltage power supply that relies on an electric field to capture particles	
Phase II: An enclosed, non-vented capillary spray system	
Phase III: Moved operation to vented space	

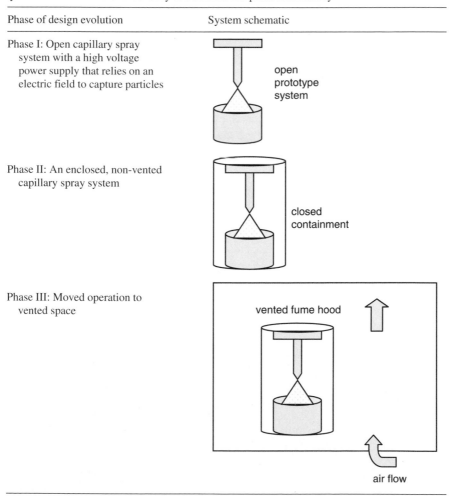

Table 1.1 (continued)

Phase of design evolution	System schematic
Phase IV: Closed chamber individually vented to external exhaust	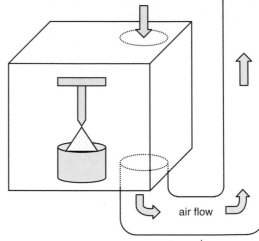
Phase V: Added filtration system to external exhaust	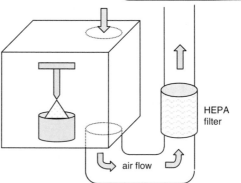
Phase VI: Modified chamber gas flow to provide and monitor maintenance sub-ambient pressure in order to maximize particle containment	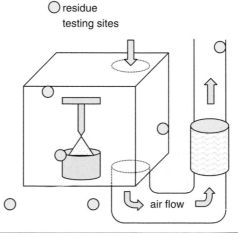

Table 1.1 (continued)

Phase of design evolution	System schematic
Phase VII: Clean room manufacturing environment; full control over gas flow through chamber and exhaust vent system, filtration removal system	maintain negative pressure in chamber — flow controls — ΔP — HEPA filter — air flow

In Phase I, the particles were produced by a simple electrospray device, consisting of a spray nozzle and platform. Since these particles were charged by the high voltage power supply, a substrate containing the opposite charge was used for collection. The collection efficiency for this process was close to 90%, as determined by the weight of material collected.

In order to capture the 10% of the particles, which were escaping the collection substrate, an enclosure was built around the spraying apparatus in Phase II. Thus collection of particles occurred on the interior surfaces and was contained in the enclosed volume. However, routine opening of the chamber potentially allowed the particles to escape. Unlike larger particles, the nanoparticles are highly mobile in air and are not influenced by settling.

As mentioned above, the spray area needs to be vented to avoid any fugitive particles. In Phase III, in order to avoid these fugitive particles, the prototype electrospray system was operated within a vented fume hood. Because this was inconvenient and restricted the amount of bench space available, the iteration in Phase IV placed the electrospray operation within a tight enclosure that was continually purged with a dilutant gas stream and the effluent stream vented directly into the fume hood system.

While the quantities being sprayed were minimal, this configuration nevertheless had the potential to release small amounts of nanoparticles into the outer atmosphere where they have the potential to remain airborne for an extended period of time. In Phase V, in order to prevent this release, a high efficiency particulate air (HEPA) filter was added which has a theoretical efficiency approaching 100% for removing the particles in these size ranges. One way to monitor the efficiency of the filter is to measure the pre- and post-filter particle concentrations in the air stream using the condensation particle counter (CPC). The CPC has the capability to monitor particle sizes down to 3 nm. Particle levels were at the threshold of detection, so the next step in Phase VI involved checking for levels of drug residue that may have become deposited on the surfaces of the system at various points. The levels of drug residue at various sites were measured by swabbing the surface of the chamber and exhaust tubing pre- and post-filter using ethanol or other solvents which are known to dissolve the drug being sprayed. The swab was then rinsed with excess solvent to collect the drug so that it could be analyzed by high precision analytical instrumentation such as HPLC. This analysis method provided the pre- and post-drug concentration measurements which confirmed for those particular experiments that the filter was operating at 100% efficiency. Periodic monitoring in pre-filter sites provides a convenient check on cleaning efficiency and potential exposure of the operator to this material (Hinds 1982).

The filter choice should be made based on its collection efficiency of the particle in the required size range (Fig. 1.2). Filtration is the most common method of capturing particles in air media (aerosol) and filters capture particles according to size. For an aerosol, the mechanisms of particle removal by the filters are diffusion, interception, inertial impaction, gravitational sedimentation, and sieving. The efficiency can be enhanced by an electrostatic effect. For nanoparticles in air, the major removable mechanism is diffusion, which affects particles less than 100 nm. As the particles traverse through the filter, they diffuse from the gas stream by Brownian motion and attach to the filter fiber. As a result, the particles are removed from the

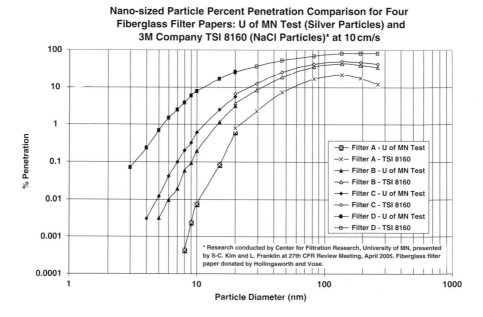

Fig. 1.2 Graph showing the filter efficiency of fiberglass filter paper produced by Hollingsworth and Vose [Graph used with the permission of Dr. David Pui]

air stream. The factors which affect particle removal are filter properties (thickness, fiber or pore diameter, diameter uniformity, solid volume fraction, degree or particle loading and electrostatic charge on the fiber), particle properties (size distribution, phase of particle (solid or liquid) and electrical charge), and gas conditions (velocity through the filter, temperature, pressure, and viscosity). The procedure for testing filter efficiency for nanoparticles is discussed by Kim et al. (2007).

Another important concern focuses on reducing the amount of exposure experienced by the operator during the electrospray operation. A further refinement to the system in Phase VII involved establishing and monitoring the presence of slight negative pressure in the spray chamber such that the interior gas should always flow away from the operator and into the filter exhaust system. The negative pressure can be monitored using a pressure gauge. This gauge monitors the pressure differential between the outside and inside of the chamber to ensure that the inner pressure is always lower than the outer pressure. This requires balancing the gas stream flow into and out of the chamber. Due to fluctuations, managing the flow through the chamber with mass flow controllers provides a convenient and monitorable safeguard. Additionally, before the chamber is opened to remove the spray sample, a dilutant gas is pumped and exhausted through the system for at least 10 volume exchanges to minimize exposure to the operator. Monitoring the particle concentrations can help to establish the number of necessary exchanges. The operator's cumulative exposure in the immediate environment of the electrospray system can also be monitored by personal environmental monitors (PEM) which are commercially available (Fig. 1.3). These monitors are placed near to the inhalation area. The PEM uses a pump to flow air through the monitor equivalent to the volume of air consumed during human breathing. This is discussed in more detail later.

With this phase complete, the transition could be made from an early R&D program to a mature operating environment with appropriate controls and monitoring potential. This system can be placed in a clean room production facility with maximum safeguards for avoiding contamination of the workspace with the nanoparticles being generated.

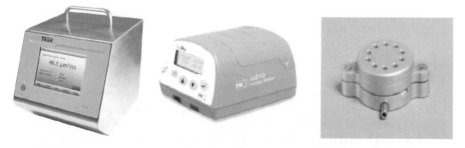

Fig. 1.3 Examples of personal environmental monitors: TSI AEROTRAK™ 9000 Nanoparticle Aerosol Monitor (*left*), TSI Model AM510 SIDEPAK™ Personal Aerosol Monitor (*middle*), and M200 Personal Environmental Monitor from MSP Corporation (*right*)

1.4 General Approaches for Categorizing Risk

Nanoparticle size and shape play an important role in the entry route and overall toxicity of a nanomaterial. Nanoparticles are produced with two or more dimensions between 1 – and 100 nm, in different shapes such as tubes, spheres, plates, fullerenes, and needles. The shape of the nanoparticle, in conjunction with the size, help determine penetration depth, as well as how long the body may require to remove the particles. Small diameter particles, such as the plates of nanoclays, will penetrate deeper into the body while larger diameter particles, such as fullerenes, will not penetrate as deeply. Long fibrous particles, such as single-walled nanotubes, are more difficult for the body to remove (Kandlikar et al. 2007).

The best example of how size and shape play a role in overall toxicity is asbestos. The features that make asbestos a hazard are related to individual fiber diameter (at <3 μm, it can be inhaled into respiratory bronchioles), length (fibers <15 μm cannot be removed by the macrophages), and its overall resistance to dissolution (Seaton 2006). Upon entry into the body, the fibers cause an inflammatory reaction that ultimately leads to fibrosis and carcinogenesis (Mossman et al. 1990).

When considering the appropriate personal protection equipment to use when performing experiments with a new material, the three most likely routes of entry into the body must be investigated: inhalation, ingestion, and dermal adsorption.

1.4.1 Inhalation

As was previously discussed for asbestos, nanoparticle size and shape play a role in the selection of deposition site, as well as, the toxicological effects resulting from the introduction of nanoparticles into the lungs via inhalation. Although there are few studies focusing on the inhalation of newer nanoparticles, carbon nanofibers, or nanotubes, obtaining data on the toxicological effects of these materials is important.

Airborne nanoparticles predominately move by convection and diffusion. Smaller particles generally deposit in the respiratory tract by diffusion (James et al. 1991). Some recent studies have shown that particle size and shape are not the only factors that play a role in determining the pulmonary toxicity of nanoparticles. It has been determined that the factors that have a role in pulmonary toxicity include, but are not limited to (Tsuji et al. 2006):

1. particle number and size distribution;
2. dose of particle to target tissue;
3. surface treatment on particles;
4. the degree to which engineered nanoparticles aggregate/agglomerate;
5. surface charge on particles;
6. particle shape and/or electrostatic attraction potential; and
7. method of particle synthesis – i.e., whether formed by gas phase or liquid phase synthesis and post-synthetic modifications, which likely influence aggregation behavior.

Inhalation and ingestion protection is provided by a number of commercially available respirators. These include particulate dust masks, as well as both half- and full-face masks. When working with just nanoparticles such as carbon nanofibers or single-walled carbon nanotubes, a particulate respirator like the P100 dust mask available from 3 M is a good choice. This respirator carries NIOSH's highest filtration efficiency rating. This mask is 99.7% effective against oil and non-oil, certain dusts, fumes, mists, radionuclides, and asbestos-containing dusts and mists (information can be found at www.3m.com).

Most commonly used half-face respirators carry a rating of N95. A list of N95 respirator distributors can be found at http://www.cdc.gov/niosh/npptl/topics/respirators/disp_part/n95list1.html. This rating signifies a respirator that filters at least 95% of airborne particles. This means that they filter out 95% of sodium chloride particles. Although numerous variations of this filter are available, it has been found that not all N95 rated filters perform the same. An experiment performed at the University of Cincinnati using two different commercially available N95 respirators showed that the penetration threshold of 5% established for these respirators can be exceeded when used against nanoparticles in the size range of ∼30–70 nm (Balazy et al. 2006). Recently, the efficiency of a variety of filter media (four different fiberglass, four different electret, and one nanofiber) was tested using silver nanoparticles. Using a nano-DMA (differential mobility analyzer), it was found that all of the filters performed at 99.99% efficiency for particle sizes down to 3 nm (Kim et al. 2007). Figure 1.4 shows commonly used particulate and half mask respirators.

1.4.2 Dermal Exposure

With an increase in the use of nanoparticles as additives, for example lotions and creams that contain nanoscale TiO_2 and ZnO, it is imperative that an understanding of the absorption potential of these materials is understood. It has been shown that nanoparticles in conjunction with motion, such as wrist movement, can penetrate the stratum corneum of human skin, reaching the epidermis and, occasionally, the dermis. This work was performed using fluorescent microspheres between 0.5 and 1.0 μm in size (Tinkle et al. 2003).

Fig. 1.4 Examples of a P100 particulate respirator (*left*) and an N-95 rated half mask respirator (*right*) available from 3 M Company

Table 1.2 Comparison of latex versus nitrile gloves

Glove type	Advantages	Disadvantages	Use against
Latex	Low cost, good physical properties, dexterity	Poor vs. oils, greases, organics. Frequently imported; may be poor quality. May result in allergic reactions	Bases, alcohols, dilute water solutions; fair vs. aldehydes, ketones, light irritant protection, infectious agents
Nitrile	Low cost, excellent physical properties, dexterity	Poor vs. benzene, methlyene chloride, trichloroethlyene, many ketones	Oils greases, aliphatic chemicals, xylene, perchlorethylene, trichlorethane, fair vs. toluene

Dermal penetration can be controlled through the use of gloves and laboratory coats. Wearing a lab coat minimizes the amount of skin surface area available for airborne nanomaterials. A lab coat will also prohibit the introduction of nano-materials onto the skin surface that results from skin contact with accidental spills. Gloves provide a barrier with the hands that also prohibit the introduction of nanomaterials. The most commonly used gloves are latex and nitrile. Table 1.2 provides a comparison of latex versus nitrile gloves in areas including advantages, disadvantages, and the appropriate type of glove to use when working with specific chemicals (information can be found at http://www.asp.anl.gov/Safety_and_Training/User_Safety/gloveselection.html).

One of the main concerns when using latex gloves is the possibility of the development of allergies. It has been discovered that certain small molecules that are used in the manufacturing process become irritants that can cause allergic contact dermatitis (ACD). One example results from the use of 2-mercaptoenzothiazole (MTB) used to accelerate polymerization during manufacturing. Researchers at the National Institute for Occupational Safety & Health (NIOSH) in conjunction with Portland State University in Oregon found that if MTB is oxidized, the resulting disulfide compound has the potential to trigger an allergic response (Fig. 1.5) (Burks 2007).

Concerns about pinholes and porous channels in either type of glove have not been borne out in careful studies. While openings occasionally can be detected, their magnitude is insufficient to provide the potential for significant exposure (e.g., Ref: Lancet publication).

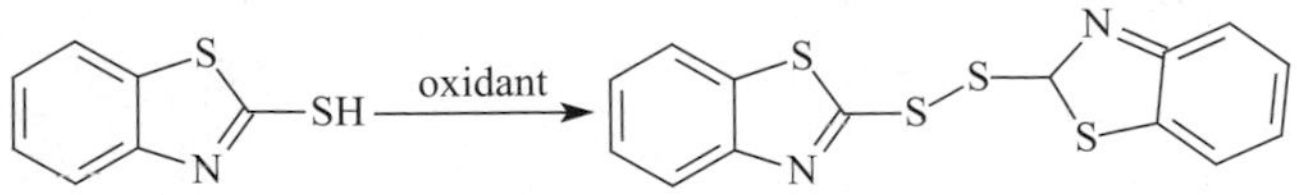

Fig. 1.5 The oxidation of MTB forms one of the disulfide components that is believed to help cause latex allergies

1.4.3 Ingestion

Ingestion of nanoparticles primarily results from hand-to-mouth contact; however, relevant research on this type of exposure in the workplace is limited at this time. In a recent review, Oberdörster et al. (2005) suggest that nanoparticles may enter the gastrointestinal tract as they are cleared from the upper respiratory tract and that they appear to pass through relatively quickly. It is uncertain what effects the nanoparticles would have on the digestive tract as they move from the oral cavity into the acidic gastric environment. It is known that some nanoparticles, specifically nanotubes and nanofibers, can undergo surface oxidation and length shortening in the presence of strong acid conditions, so it is possible that the acidic gastric environment will not only be unable to destroy the nanoparticles, but it may functionalize and shorten the particles to a point that their toxicological effect on the surrounding environment might change drastically.

Minimizing ingestion of nanomaterials can be accomplished by maintaining a clean work environment, wearing a respirator or particulate mask, and washing exposed areas such as hands and arms.

1.5 Regulatory Aspects of Working with Nanoparticles

Current regulatory oversight that is specifically directed toward nanoparticles is limited, primarily addressing nanoparticles when they fall within a specific size range (as in fine or ultrafine dust particles) or composition, including heavy metals or some other compound with known toxicity. Materials used in the manufacturing of nanoparticles, such as solvents, are governed by toxic waste regulations.

For most small companies and university laboratories, encounters with regulatory bodies will come as a matter of following standard laboratory safety practices, radiation license requirements, and toxic waste handling. For example, our initial regulatory contact occurred within the setting of a university small business incubator on the University of Minnesota campus. Because the laboratory was located within the University environment, the company was required to submit a laboratory safety plan in accordance with a standard university template. Periodically, unannounced laboratory visits were made by county hazardous waste officials as well as campus radiation safety officers. The infrastructure that was supplied within the university was convenient and required almost no independent policies or procedures to be created.

Upon our move to an outside facility (the former state crime lab building), we were fortunate in securing space with fully outfitted laboratory fume hoods, fire blankets, eye wash stations, and other standard safety equipment. The second regulatory contact came shortly after our move in the form of an unscheduled visit by the county public health department's hazardous waste division inspector. This was an informative, helpful meeting in which our laboratory facilities were inspected and guidelines and practices for managing hazardous waste as well as an application for the appropriate hazardous waste license were provided. A visit to the county website reveals standard, EPA-conforming practices that are based on the quantity of hazardous waste generated by the business. Repeat inspector visits have occurred infrequently, and always unannounced, with inspections related specifically to the items identified in the initial visit.

1.5.1 Hazardous Waste Management Regulations

Waste generators are classified as very small quantity generators, small quantity generators, and large quantity generators, as per the following criteria listed in Table 1.3. We were categorized as a very small quantity generator during our initial pre-license visit.

The county provides guidelines and requirements for the following hazardous waste topics, (on the next page) with specific guidance that relates to the quantity of waste generated.

Table 1.3 Categories of hazardous waste generators as implemented by Public Health Department in Ramsey County, Minnesota. Hazardous waste information obtained at http://www.co.ramsey.mn.us/ph/hw/hw_mangement.htm#getting_licensed

Category	Criteria
Very Small Quantity Generators (VSQG)	• Generates 100 kg or less of waste per month (less than $\frac{1}{2}$ drum). • Can accumulate 1000 kg (or about 4 drums) on-site. If a generator exceeds this limit, waste must be managed according to Small Quantity Generator guidelines. • Can store less than 1000 kg indefinitely. Once 1000 kg is accumulated, waste must be shipped off-site within 180 days.
Small Quantity Generator (SQG)	• Generates between 100 and 1000 kg of waste per month (about 1/2 to 4 drums). • Can accumulate 3000 kg (or about 12 drums) of waste on-site. If a generator exceeds this limit, waste must be managed according to Large Quantity Generator guidelines. • Waste must be shipped off-site within 180 days after the waste was first placed in the container.
Large Quantity Generator (LQG)	• Generate 1000 kg or more of waste per month (more than 4 drums) • No limit on how much waste can be accumulated on-site. • Waste must be shipped off-site within 90 days after the waste was first placed in the container.

- Indoor storage
- Outdoor storage
- Labeling of containers
- Labeling for transport
- Criteria for selecting a transport service
- Determining the frequency of pick-up
- Choosing a hazardous waste facility
- Emergency management planning, response and equipment
- Management of spill clean-up and notification requirements
- Personnel training, depending on quantities generated
- Record keeping requirements and recommendations and notification require-
  ments.

Setting up conforming hazardous waste management practices is time well spent, particularly given the unexpected controversy that may develop around the topic of nanotechnology and the environment or human health and safety. Not all municipalities may have requirements as detailed as those for a large metropolitan area, but the guidance that is supplied is useful regardless of a company's location. Not only complying with local regulations, but also exceeding these requirements, is a good risk management practice for a start-up.

1.5.2 Evolving Regulations

Existing U.S. regulations do not specifically address nanomaterials. Certain regulations and associated governmental agencies apply, such as those mentioned above that govern the use and disposal of hazardous bulk materials used in the manufacture of nanomaterials or those that govern workplace exposure to airborne fine particulate matter. These are discussed in more detail in the following sections.

1.5.2.1 EPA/Toxic Substances Control Act

The Toxic Substances Control Act (TSCA) was enacted by Congress in 1976. This act enables the Environmental Protection Agency (EPA) to track the 75,000 (at that time) industrial chemicals that were produced or imported into the USA. This also allows the EPA to implement tracking mechanisms and controls for the thousands of chemicals produced each year with either unknown or dangerous characteristics (http://www.epa.gov/Region5/defs/html/tsca/htm). While TSCA does not specifically apply to nanomaterials, it does provide the EPA with responsibility for evaluating "new chemicals" or "significant new uses" of existing chemicals. Assuming that nanomaterials may fall under the new chemical or significant new use category, there are reporting requirements and a series of exemptions that may be applied for by anyone using these types of materials. In October 2006, the EPA initiated an outreach program to arrive at an approach for managing nanomaterials, which resulted in the publication in July 2007 of the "Concept Paper for the Nanoscale Materials Stewardship Program under TSCA."

1.5.2.2 Occupational Safety and Health Administration (OSHA)

Occupational Safety and Health Administration sets enforceable permissible exposure limits (PELs) standards to protect workers against the health effects of exposure to hazardous substances. For example, in the case of asbestos, employee exposure must not exceed 0.1 fiber per cubic centimeter (f/cc) of air, averaged over an 8-hr work shift. Short-term exposure must also be limited to not more than 1 f/cc, averaged over 30 min. PELs are regulatory limits on the concentration of a substance in the air. They may also contain a skin designation. PELs are contained in 29 CFR 1910.1000, the air contaminants standard. OSHA standards can be found at http://www.osha.gov. OSHA PELs are based on an 8-hr time weighted average (TWA) exposure.

The American Conference of Governmental Industrial Hygienists (ACGIH) has also provided related non-regulation guidelines for the exposure of employees to airborne concentrations of particles and chemical substances. ACGIH uses the threshold limit value (TLV) which is defined as the reasonable level to which a worker can be exposed day after day without adverse health effects. TLVs (along with biological exposure indices or BEIs) are published annually by the ACGIH (ACGIH 1986).

The TLV is an estimate based on information gathered from industrial experience, from experimental human or animal studies of a given chemical substance, and the reliability and accuracy of the latest sampling and analytical methods. The values can be adjusted based on new research and can often modify the risk assessment of substances while new laboratory or instrumental analysis methods can improve analytical detection limits. The TLV is a recommendation by the ACGIH with only a guideline status.

The National Institute of Occupational Safety and Health (NIOSH) publishes recommended exposure limits (RELs) which OSHA takes into consideration when promulgating new regulatory exposure limits. Also NIOSH is conducting research on the occupational safety and health implications and applications of nanotechnology (information available at www.niosh.gov). In October 2005, NIOSH issued a draft report entitled "Approaches to Safe Nanotechnology," which was updated again in October 2006. This report identified potential health and safety concerns associated with nanomaterials, risks of exposure in the workplace, approaches for assessing workplace exposure, and precautionary measures to be employed when risks are unknown or while they are being quantified.

1.5.2.3 Recent Developments

The regulatory environment is rapidly evolving. The government issued reports mentioned in this section are recommended reading for individuals working with nanomaterials. Other U.S. governmental agencies and interagency working groups are examining the need for such regulations. For example, the Food and Drug Administration released a report entitled "Nanotechnology: A Report of the U.S. Food and Drug Administration Task Force" in July 2007, which identified FDA's assessment of the emerging field of nanoscale materials and their interaction with biological systems. No regulations were proposed in the report, but the FDA offered guidelines to manufacturers to consider and suggested consultation with the Agency early in the development of products using nanoscale materials. Interestingly, the report discussed whether FDA is obligated to consider the environmental impact of nanoscale materials in an FDA-regulated product, under the National Environmental Policy Act. Separately, under direction of the National Science and Technology Council, the interagency Nanotechnology Environmental and Health Implications Working Group issued a report in August 2007 highlighting the research needs for considering the environmental, health and safety issues related to engineered nanoscale materials [Information can be found on the Nanotechnology Environmental and Health Implications Working Group link located in reference section].

Occasionally, due to local concerns, a municipality may enact regulations that exceed those of state or national agencies. For example, the city council of Berkeley, California enacted the Hazardous Waste Amendment to the Hazardous Materials Ordinance in December, 2006, which provides for specific reporting requirements for manufactured nanoparticles, toxicity information, if available, and handling, monitoring, containment, and disposal practices (Monica et al. 2007). Notably, in the Berkeley ordinance there is no lower limit on material quantities that meet the requirements for disclosure.

Companies are well-advised to follow developments on the local, national, and international fronts, because proactive management of risk and record-keeping is always easier than playing catch-up and possibly slowing the pace of development.

1.6 Conclusions

Establishing a safe environment for working with nanoparticles technology must necessarily be described as an evolutionary process. Thorough review of the basic toxicity potential of all reagents and materials at the time they are acquired by a facility is a first step. Ongoing assessment of the nature of particulate matter that could be released into the work environment will need to be done on a case specific basis, using analytical equipment that is either available on site or with regular monitoring by personal exposure sampling, where samples are analyzed off site.

The regulatory setting is also evolving. Most regulatory agencies in the USA and in Europe have released position papers that discuss how they are approaching regulation of nanoparticle technology; how these will be translated into regulations is not predictable at the present time. Small enterprises are advised to follow these processes closely and to comment where their unique expertise may provide valuable inputs into the regulatory process.

Acknowledgments We should acknowledge the guidance and support of our two inventors, David Pui and Da-Ren Chen, whose expertise in nanoparticles aerosol behavior has been invaluable during the evolution of our spray equipment system. We also thank Jennifer Kuzma for her insights and the opportunity to participate in a forward looking exercise on nanotechnology regulation at the Hubert H. Humphrey Institute of Public Affairs, University of Minnesota.

References

American Conference of Governmental Industrial Hygienists (1986) Industrial ventilation: a manual of recommended practice, 19th edn. Edward Brothers Incorporated, Ann Arbor, MI.

Balazy A, Toivola M, Reponen R, Podgorski A, Zimmer A, Grinshpun SA (2006) Manikin-based performance evaluation of N95 filtering-facepiece respirators challenged with nanoparticles. Ann Occup Hyg 50(3):259–269.

Burks R (2007) Key Step Found in Latex Allergy, C & EN, July 30, 16.

Chen D-R, Pui DYH (1997) Experimental investigation of scaling laws for electrospray: dielectric constant effect. Aerosol Science Technol 27:367–380.

Chen D-R, Pui DYH (2000) Electrospraying apparatus and method for introducing material into cells. U.S. Patent No. 6,093,557.

Environmental Protection Agency (2007) Concept paper for the nanoscale materials stewardship program under TSCA. http://www.epa.gov/opptintr/nano/nmspfr.htm

Hinds WC (1982) Aerosol technology: properties, behavior, and measurement of airborne particles. John Wiley & Sons, Inc., New York, NY.

James AC, Stahlofen W, Rudolf G, Egan MJ, Nixon W, Gehr P, Briant JK (1991) The respiratory tract deposition model proposed by the ICRP Task Group. Radiat Prot Dosimet 38:159–165.

Kandlikar M, Ramachandran G, Maynard A, Murdock B, Toscano WA (2007) Health risk assessment for nanoparticles: a case of using expert judgment. J Nanoparticle Res 9:137–156.

Kim SC, Harrington MS, Pui DYH (2007) Experimental study of nanoparticle penetration through commercial filter media. J Nanoparticle Res 9:117–125.

Monica JC, Heinz ME, Lewis PT (2007) The perils of pre-emptive regulation. Nat Nanotechnol 2:68–70.

Mossman BT, Bignon J, Corn M, Seaton A, Gee JBL (1990) Asbestos: scientific developments and implications for public policy. Science 247:294 301.

Nanotechnology Environmental and Health Implications Working Group (2007) Prioritization of environmental, health and safety research needs for engineered nanoscale materials: an interim document for public comment. http://www.nano.gov/html/society/ehs_priorities/

National Institute for Occupational Safety and Health (2006) Approaches to safe nanotechnology: an information exchange with NIOSH. http://www.cdc.gov/niosh/topics/nanotech/safenano/

Oberdörster G, Oberdörster E, Oberdörster J (2005) Nanotoxicology: an emerging discipline evolving from studies of ultrafine particles. Environ Health Perspect 113:823–839.

Seaton A (2006) Nanotechnology and the occupational physician. Occup Med 56:312–316.

Tinkle SS, Antonini JM, Rich BA, Roberts JR, Salmen R, DePree K, Adkins EJ (2003) Skin as a route of exposure and sensitization in chronic beryllium disease. Environ Health Perspect 111:1202–1208.

Tsuji JS, Maynard AD, Howard PC, James JT, Lam C-W, Warheit DB, Santamarial AB (2006) Research strategies for safety evaluation of nanomaterials, Part IV: Risk assessment of nanoparticles. Toxicology 89(1):42–50.

U.S. Food and Drug Administration (2007) Nanotechnology: a report of the U.S. Food and Drug Administration Task Force. http://www.fda.gov/nanotechnology/nano_tf.html

Chapter 2
Cytotoxicity of Photoactive Nanoparticles

Yuhui Jin and Xiaojun Zhao

Abstract This chapter describes the cytotoxicity of photoactive materials (specifically, quantum dots, noble metal nanoparticles (including gold and silver), and fluorescent silica nanoparticles). A thorough representation of in vitro and in vivo toxicity studies is presented. Since the toxicity on photoactive nanomaterials described in this chapter has developed rapidly and has attracted a great amount of interest, it is expected that many novel developments and applications of photoactive nanomaterials will ensue in the near future.

Abbreviations

Ag NPs:	Silver nanoparticles;
Au NPs:	Gold nanoparticles;
BSA:	Bovine serum albumin;
CdSe:	Cadmium selenide;
CdTe:	Cadmium telluride;
EC_{50}:	Effective concentration;
GSH:	Glutathione;
LC_{50}:	Lethal concentration;
MMP:	Mitochondrial membrane potential;
MTT:	3-(4,5-dimethylthiazol-2-yl)-2,5-diphenyltetrazolium bromide;
MPA:	Mercaptopropionic acid;
MUA:	Mercapto-undecanoic acid;
NAC:	N-Acetylcysteine;
NPs:	Nanoparticles;
PEG:	Polyethylene glycol;
QDs:	Quantum dots;
ROS:	Reactive oxygen species;

Y. Jin (✉)
Department of Chemistry, University of North Dakota, Grand Forks, ND 58202, USA
e-mail: yuhui.jin@und.nodak.edu

T.J. Webster (ed.), *Safety of Nanoparticles*, Nanostructure Science and Technology,
DOI 10.1007/978-0-387-78608-7_2, © Springer Science+Business Media, LLC 2009

Silica NPs: Silica nanoparticles;
    SOPC: Phosphotidylcholine;
    SOPS: Phosphotidylserine;
      ZnS: Zinc sulfide.

Contents

2.1 Introduction

Photoactive nanomaterials are one of the most important nanomaterials for functioning as highly sensitive labeling reagents in various bioapplications (Cao et al. 2001; Dubertret et al. 2002; Zhao et al. 2003a, 2004; Liang et al. 2005). Current developments have demonstrated that photoactive nanomaterials provide significant strong and photostable optical signals reporting the presence of trace amounts of analytes. In comparison with traditional optical labeling techniques, which usually use fluorescent molecules to signal target analytes, the photoactive nanoparticle enhances detectable signals 10–10000 times (Mirkin et al. 1996). So far, these nanomaterials have been used in the detection of trace amounts of DNA, mRNA, and proteins (Cao et al. 2002, 2003; Lagerholm et al. 2004). Most importantly, the photoactive nanomaterials have demonstrated great potential as highly efficient labeling regents for in vitro and in vivo study of various cells (Michalet et al. 2005; Santra et al. 2005; Yi et al. 2005).

In the application of photoactive nanomaterials with living cells, a major concern, is whether these nanomaterials would cause toxic effects to living systems. Recently, some scientists have started working on the cytotoxicity of photoactive nanomaterials. Although current cytotoxicity studies are at the initial stage, results are significant for the further development and application of photoactive nanomaterials in the biological and biomedical fields.

In this chapter, some recent cytotoxicity studies on several typical photoactive nanomaterials will be reviewed. These nanomaterials include quantum dots (QDs), noble metal nanoparticles (gold nanoparticles (Au NPs) and silver nanoparticles

(Ag NPs)), and fluorescent silica nanoparticles (Silica NPs). Various factors that affect the cytotoxic properties of nanomaterials will be discussed briefly, such as the release of metal ions and particle size. Meanwhile, the general strategies of reducing/eliminating the cytotoxicity of these nanomaterials will be summarized in this chapter as well.

2.2 Quantum Dots

Quantum dots (QDs) or semiconductor nanocrystals are considered to be one of the best substitutes for conventional organic fluorescent materials. Currently, the most widely used QDs are made from cadmium and selenium. The diameter of the QDs is around several nanometers. Because of their composition and size, QDs are endowed with certain physically tailorable chemical properties, such as a tunable fluorescence emission wavelength and a strong resistance against photobleaching. Therefore, QDs have been recruited as fluorescent probes in biological and biochemical studies.

Although cadmium selenide quantum dots (CdSe QDs) seem to be the next generation of fluorescent materials for medical uses, their toxicity might be a major concern. Since cadmium is an acutely toxic heavy metal even at very low concentrations, the potential toxicity properties of Cd QDs should be evaluated prior to their real applications in the field of biomedicine and biology. So far, a few literature articles have reported the cytotoxic effects and DNA damage of Cd QDs (CdSe QDs and cadmium telluride (CdTe) QDs). The toxic properties of other heavy metal QDs, for example, PbS QDs, have not been reported yet. Thus, this chapter will only focus on the cytotoxicity of cadmium QDs and the major factors that cause such toxic effects.

2.2.1 Cytotoxic Effects of Cd QDs

2.2.1.1 Release of Cd^{2+} Ions

The toxicity of Cd QDs was first studied by Derfus et al. (2004). Primary hepatocytes were incubated with CdSe QDs for 24 hrs following a measurement of cell mitochondrial activity through an MTT viability assay. Severe cytotoxic effects of CdSe QDs were observed even at a low concentration of 62.5 $\mu g \cdot mL^{-1}$ of QDs under certain conditions, such as perturbing the CdSe QDs with UV-light, oxygen or hydrogen peroxide. The release of free Cd^{2+} ions from CdSe QDs after surface oxidation was a key reason for their cytotoxicity. The experiments clearly demonstrated that the extent of cytotoxic effects was correlated with the concentration of the released Cd^{2+}.

To reduce the cytotoxic effect from the surface oxidation, proper postcoating of CdSe QDs was an effective solution. In the effort of achieving this goal, zinc sulfide (ZnS) and bovine serum albumin (BSA) surface coated CdSe QDs were developed. The ZnS- and BSA-coated CdSe QDs showed a significant decrease of their cytotoxicity, but it was not completely eliminated. Kirchner et al. (2005) further proved this hypothesis by testing the detachment of NRK fibroblasts from the cell culture substrate upon incubation of the cells with QDs. They found the stability of the postcoating shell of the CdSe QDs was crucial to the QDs cytotoxicity. A stable and well-covered shell reduced the release of Cd^{2+} dramatically. For example, the ZnS shell of CdSe/ZnS QDs increased the critical concentration (no toxic effects were observed at this concentration of QDs) by almost a factor of 10. The effect of a silica shell on the QDs was also investigated. The results demonstrated that a stable crosslinked silica shell successfully reduced the cytotoxic effects of QDs. Furthermore, as larger polymer compounds were employed for the second postcoating silica-coated QDs, the toxic effect of QDs became undetectable. At concentrations of 30 μM of surface atoms of Cd, both polyethylene glycol (PEG) silica-coated CdSe and CdSe/ZnS QDs showed no toxic effects.

Lovrić et al. (2005) found a similar cytotoxic effect of CdTe QDs as that of CdSe QDs. With the hypothesis that free Cd^{2+} ions was a major cause for their cytotoxicity, two antioxidants, N-Acetylcysteine (NAC) and Trolox, were employed to pretreat P12 cells for 2 hrs. Based on previous studies, NAC and Trolox protected cells against Cd^{2+} induced cell death. The QD-induced reduction in cell metabolic activity completely disappeared in the NAC pretreated cells. However, Trolox showed no improvement regarding the QD-induced cytotoxicity. This result suggested that the release of free Cd^{2+} ions was one major factor, but not the only cause for cytotoxicty. Lovrić et al. (2005) listed three possible mechanisms that NAC reduced cytotoxicity: (a) Stabilized the QDs in the media by being absorbed onto the QDs' surface; (b) Enhanced glutathione (GSH) expression of cells to prevent QD-induced cytotoxicity; and (c) Activated key antiapoptotic signal transduction pathways that lead to transcription of genes involved in cell survival. The stabilization function of the postcoating method has also been demonstrated in BSA-coated CdTe QDs.

2.2.1.2 Chemical Properties of Surface Molecules

In addition to the release of free Cd^{2+} ions from Cd QDs, some surface-covering molecules have also contributed to the cytotoxic effects of QDs. Hoshino et al. (2004) investigated the cytotoxic effects of CdSe/ZnS QDs with different surface-covering groups using the comet assay, flow cytometry, and the MTT viability assay. WTK1 cells and Vero cells were employed as target cells to incubate with CdSe/ZnS QDs that were coated with mercapto-undecanoic acid (MUA) (QD-COOH), cysteamine (QD-NH$_2$), or thioglycerol (QD-OH) groups. After 2 hrs of incubation of the cells with the QDs, the MUA-coated QDs (QD-COOH) showed severe cytotoxicity at doses greater than 100 μg·mL^{-1} of QDs, while the thioglycerol-coated QDs showed slightly cytotoxic effects to the target cells. These three groups of QDs, which have the same core composition but different surface chemical molecules, demonstrated obvious different cytotoxic effects. Based on this result, Hoshino et al. concluded that the chemical properties of surface-covering molecules on the QDs affected the toxic effects of QDs significantly.

2.2.1.3 Size Effect

The different sizes of QDs have showed different cytotoxic effects. Shiohara et al. (2004) compared the cytotoxic effects of three different sized QDs: QD520 (green fluorescence QD), QD570 (yellow fluorescence QD), and QD640 (red fluorescence QD). QD520, which has the smallest size among the three QDs, showed the highest extent of cytotoxic effects. The explanation for the size effect was that the mobility of the QDs inside the cells depends on its size. The small size gave QDs a better mobility than the larger ones. Thus, the small QDs had more of a chance to contact cells and further caused a higher extent of cytotoxicity.

The size effect was also observed during a cytotoxicity study of CdTe QDs. Lovrić et al. (2005) found the green fluorescence CdTe QDs (2.2±0.1 nm in diameter) exhibited higher cytotoxic effects than the red fluorescence QDs (5.2±0.1 nm in diameter). The green CdTe QDs formed stable colloids in solution for more than one month, while the red ones aggregated easily. The actual size of the red fluorescence QDs might be larger than 5.2 nm. Using a fluorescence microscope, the interaction of the CdTe QDs with cells was monitored in situ. The green fluorescence QDs penetrated in the nucleus membranes of N9 cells and remained inside the nucleus after 1 hr of incubation. At the same conditions, the red fluorescence QDs stayed in the cytosol, but could not enter the nucleus. Interestingly, when the green fluorescence QDs were post-coated with macrobiomolecules such as BSA, these larger sized QDs were not able to enter the nucleus. The results suggested that the cytotoxicity of QDs was reduced by increasing particles size.

2.2.1.4 Other Effects

The stability of coating ligands on the QDs affected their cytotoxicity. Kirchner et al. (2005) investigated different compounds coated on CdSe QDs, in which binding forces between QDs and coating ligands were different. The results showed that less stable polymer-coated CdSe QDs and CdSe/ZnS QDs had larger cytotoxic effects than the corresponding mercaptopropionic acid (MPA)-coated QDs. The previous experiments confirmed that MPA was immobilized in a stable manner onto CdSe QDs. Thus, to reduce the cytotoxicity of QDs, the improvement of coating ligand stability on QDs is an effective approach.

In addition, the extent of QD cytotoxicity was related to the type of cell. Shiohara et al. (2004) tested the cytotoxic effects of MUA-QDs (CdSe QDs) to three cell lines – Vero cells (African green monkey's kidney cells), Hela cells, and primary human hepatocytes. The QDs caused much less damage to Vero cells than the other two cell lines at the same experimental conditions.

2.2.2 DNA Damage by QDs

It has been reported that some QDs cause DNA damage. Water soluble CdSe/ZnS QDs damage super-coiled double-stranded DNA through DNA nicking through incubation of the QDs with the DNA. The nicking effect resulted in the breaking of deoxyribose units and the uncoiling of the double strands of DNA. Green and Howman (2005) discovered this phenomenon using a plasmid nicking assay. Fifty-six percent of DNA was damaged after 1 hr of incubation with MPA-coated CdSe/ZnS QDs under UV light. Meanwhile, in the same condition, the control samples without QDs showed only 5% of DNA damage. Furthermore, when the cell samples were treated with QDs in the dark (no UV radiation), 29% of DNA strands were damaged after 1 hr of incubation. Green et al. concluded that the obvious DNA damage was caused by free radicals released during oxidation of QDs. The ESR spectrum confirmed this hypothesis when comparing the QDs before and after radiation. An increase of the free radical signals was determined after radiation of QDs indicating more free radicals were formed by QDs. Thus, the oxidation reaction contributed to DNA damage when using QDs.

In summary, some QDs have shown significant toxic effects to cells and DNA strands. These effects might be a significant obstacle for the further applications of QDs in the field of biomedicine. To reduce the toxic effects of QDs, proper protections like postcoating QDs should be employed.

2.3 Noble Metal Nanoparticles

Nanostructures made from noble metals, Au or Ag, have been employed as photoactive labels in a variety of biosensing systems. The noble metal NPs have strong, size-dependent optical properties and UV-visible extinction bands. Thus, different sized metal NPs give different colors. From this point of view, metal NPs have similar properties with QDs with the major exception that their nominal size is much bigger than QDs. However, compared to QDs, gold and silver nanoparticles have several advantages, particularly their ease of synthesis and ability to bind to various target molecules. So far, gold nanoparticles (Au NPs) and silver nanoparticles (Ag NPs) have been developed and used for a wide variety of ultrasensitive chemical and biological analyses.

To explore the safety issue of gold and silver nanoparticles, the cytotoxic effects of silver and gold nanoparticles were studied recently. The effects of dosage, surface molecule properties, and sizes were investigated. The cytotoxic effects of these noble metal nanoparticles were much less than QDs, but detectable in certain situations. Some resolutions for the elimination of Au/Ag NPs toxic effects will be summarized in this section.

2.3.1 Silver Nanoparticles

2.3.1.1 Oxidative Stress

Reactive oxygen species (ROS) damage DNA and RNA strands in cells through participation in the cell apoptosis processes. Due to the redox reactive property of noble metals, Ag NPs increased the concentration of ROS, and thus caused cell death. This phenomenon was observed by Hussain et al. (2005). In the concentration range of 5–50 μg·mL^{-1} of silver nanoparticles (Ag NPs), a significant cytotoxic effect of Ag NPs (15 and 100 nm) was determined using the MTT assay after 24 hrs of incubation of Ag NPs with BRL 3A rat liver cells. This result was further confirmed by the membrane leakage of lactate dehydrogenase (LDH) assay for the concentration range of 10–50 μg·mL^{-1} Ag NPs. To address the cytotoxicity rationale, the ROS levels in cells were investigated at different time periods. The maximum ROS level was determined at 6 hrs of treatment of the cells with the Ag NPs. The ROS amounts were increased by 10 fold in the cells after incubating them with 25 μg·mL^{-1} silver nanoparticles. After 24 hrs of treating cells with Ag NPs, the mitochondrial membrane potential (MMP) of these BRL 3A cells significantly decreased (80%), and GSH was also reduced (70%). Therefore, the oxidative stress was an important cause for silver NP induced cytotoxicity. Braydich-Stolle et al. (2005) reported similar silver NP induced cytotoxicity in mammalian germline stem cells (mouse C18-4 cells). The Ag NPs reduced mitochondrial function and increased the membrane leakage. However, some mechanisms are still unknown, such as how these Ag NPs deplete GSH levels and increase ROS concentration.

2.3.1.2 Size Effect

Ag NPs have demonstrated an opposite size effect with QDs. As the size of Ag NPs decreased, the toxic effect as decreased as well. Hussain et al. (2005) found that the 100 nm Ag NPs exhibited higher cytotoxicity than that of 15 nm Ag nanoparticles. In the LDH assay, the EC$_{50}$ (effective concentration) value of 100 nm silver NPs was 24±9.25 μg·mL^{-1}. The data was much lower than that of the EC$_{50}$ of 15 nm NPs (50±10.25 μg·mL^{-1}). However, the differences of EC$_{50}$ values between 15 and 100 nm Ag NPs in the MTT assay were much smaller (100 nm Ag NPs: 19±5.2 μg·mL^{-1}, 15 nm Ag NPs: 24±7.25 μg·mL^{-1}). So far, the mechanism of the size effect is not clear.

2.3.1.3 Other Effects

The cytotoxicity of Ag NPs was slightly different depending on cell lines. Braydich-Stolle et al. (2005) compared the cytotoxic effects between C18-4 cells and BRL 3A cells. The LDH EC$_{50}$ of Ag NPs in C18-4 cells was 0.25 μg·mL^{-1}. In contrast, the LDH EC$_{50}$ of Ag NPs in BRL 3A cells was 50 μg·mL^{-1}; therefore, C18-4 cells were more sensitive than BRL 3A cells to 15 nm Ag NPs. Nevertheless, the treatment conditions were not identical (C18-4 cells: 48 hrs treatment, BRL 3A cells: 24 hrs treatment).

Until now, the effect of surface molecules on Ag NPs cytotoxic properties has not been reported. However, a study of surface molecules on Au NPs has been carried out as reviewed below. The results provided fundamental information that might be relevant to Ag NPs as well.

2.3.2 Gold Nanoparticles

2.3.2.1 Effect of Surface Modification

Pure gold nanoparticles (Au NPs) have little toxic effects due to the nature of inert elements. However, surface modification of NPs is necessary to obtain certain properties for nanoparticle applications. The surface molecules give nanoparticles' additional properties that might result in cytotoxicity of the nanoparticles. Goodman et al. (2004) compared the toxicity of cationic and anionic functionalized Au NPs (diameter of core nanoparticles was 2 nm) to Cos-1 cells by determining cell LC_{50} values (Lethal Concentration). Results showed that the cationic molecule covered Au NPs exhibited higher cytotoxic effects than that of the anionic modified molecules. After incubation of the cells with cationic molecules coated Au NPs for 1 hr, the LC_{50} value of Cos-1 cells reached 1.0 ± 0.5 μM. The LC_{50} value of the cells incubated with anionic coated Au NPs was greater than 7.37 μM after 24 hrs of incubation. To further confirm the result, red blood cells and *Escherichia coli (E. coli)* bacterial cells were tested using the same method. The results were the same as that of Cos-1 cells. The difference in cytotoxicity between the two coated Au NPs resulted from the different extent of cell membrane adhesion or cell lysis caused by the NPs. Due to the negative charges in all cells, the cationic molecule coated NPs had strong electrostatic attractions with the cells. As a result of this attraction, the NPs were drawn into the cell membranes. The hypothesis was proved by a vesicle-disruption assay. Two vesicles, SOPC (phosphotidylcholine) and SOPS (phosphotidylserine), were used in the assay. SOPS was an overall negatively charged vesicle and SOPC was a neutral vesicle. The results showed that the cationic NPs lysed negative charged SOPS more efficiently than the anionic NPs. Meanwhile, the neutrally charged SOPC showed a reversed trend for anionic and cationic nanoparticles. The assay confirmed that different surface charges of Au NPs resulted in different cell lysing efficiencies.

2.3.2.2 Effects of Cell Types and Stability of the Surface Molecules

Cell types and the stability of coating ligands on the NPs also affected the cytotoxicity of Au NPs. Goodman et al. (2004) determined the cytotoxicity of Au NPs towards two types of cells — Cos-1 cells and *E coli* bacterial cells at the same conditions. The LC_{50} value of Cos-1 cell was 1.0 ± 0.5 μM after 1 hr of incubation with cationic Au NPs, while the *E coli* bacterial cells showed a 2- to 3-fold increase in LC_{50} value. The possible explanation for this difference was that cell wall surrounding the *E. coli* bacterial cells protects the cell against the penetration of NPs. Thus, a higher concentration of Au NPs was needed to fully rupture the bacterial cells.

The stability of coating ligands on NPs affected its cytotoxicity. As described in the section of Ag NPs, stable surface ligands reduced NPs toxic effects. Kirchner et al. (2005) coated Au NPs with an inert polymer. The cytotoxic effects of the Au NPs with such stable surface ligands were much lower than that of Au NPs.

2.4 Fluorescent Silica Nanoparticles

Fluorescent nanoparticles provide highly luminescent signals due to the relatively high quantum yield of dye molecules doped inside the nanoparticles. Various organic dye-doped polymer microparticles have been developed (Ito et al. 2001; Kwon et al. 2002; Zhou et al. 2002). However, so far, little toxicity studies on this type of nanoparticle have been reported. Silica-based fluorescent nanoparticles have been rapidly developed in recent years. Tan's group in the University of Florida has made use of the fluorescent silica nanoparticles for a wide variety of applications, including the biological field (Zhao et al. 2003b; Wang et al. 2005; Chang et al. 2005). The fluorescent silica nanoparticle consists of thousands of dye molecules in a silica matrix. Due to such a large number of dye molecules, the dye-doped silica nanoparticles provide highly luminescent signals when used as optical probes. Recent applications of the fluorescent nanoparticles have demonstrated a great potential for the NPs towards becoming a revolutionary labeling materials for bioanalysis. Therefore, an investigation of the cytotoxic properties of the fluorescent silica nanoparticles is critical to direct further applications of the fluorescent NPs in the field of biomedicine.

When comparing the compositions and structures of QDs/noble metal nanoparticles to that of the fluorescent silica nanoparticles, fluorescent silica NPs contain no redox active metal atoms, which have proved to be a major cause of the cytotoxic effects of QDs (cadmium) and Ag NPs (silver). Based on the current initial research, it seemed like the fluorescent silica NPs exhibited a much lower toxicity than QDs and noble metal NPs as summarized below.

2.4.1 Low Cytotoxicity of Silica Nanoparticles

Li et al. (2002) investigated the cytotoxicity of fluorescent silica NPs to COS-7 cells using MTT assays. After being treated with the silica NPs (50 nm in diameter), COS-7 cells showed no apparent cytotoxic effects as the nanoparticle concentration was lower than 1560 $\mu g \cdot mL^{-1}$. As the concentration of silica NPs was over 1560 $\mu g \cdot mL^{-1}$, the number of living cells decreased drastically. Luo et al. (2004) also observed the low cytotoxic property of silica NPs through an MTT assay. They studied both pure silica nanoparticles and superfect contained silica NPs. The results showed that pure silica nanoparticles had no effect on cell proliferation; but the superfect contained silica NPs (concentration of NPs: 2×10^8 NPs mL^{-1}, superfect: 7.5 μg) inhibited cell proliferation about 30% after a 2 hr incubation of the cells with the NPs.

Recently our group has studied the cytotoxicity of carboxyl group coated fluorescent silica NPs to Mouse A549 cells using various bioassays. The cells were incubated with the NPs over a period of 72 hrs. The results exhibited little cytotoxicity of silica NPs to the living cells. Compared to the negative control samples (without fluorescent silica NPs), the nanoparticle treated cells retained similar percentages of survival cells (60%). These preliminary results have demonstrated a

great potential of using the fluorescent silica nanoparticles in the biomedical field
with low cytotoxicity.

2.4.2 Protection of DNA from Cleavage

In addition to the lower cytotoxic effect, fluorescent silica nanoparticles have
demonstrated a unique property – protection of DNA strands from cleavage. He
et al. (2003) reported that amino-modified silica NPs (diameter: 45±4 nm) efficiently protected plasmid DNA from enzymatic cleavage. They first investigated
the cleavage effect of DNA was cleavage enzyme – Dnase I on two groups of DNA
strands. One of the DNA strands incubated with the nanoparticles prior to the reaction with Dnase I. The results showed that the enzyme could not cleave this group
of DNA strands. However, the other group of DNA strands without incubation with
NPs were cleaved to many small strands. It seemed that the nanoparticles protected
the DNA from cleavage. To confirm the results, they further tested the Dnase I
cleavage effect on plasmid DNA. The nanoparticle conjugated plasmid DNA was
still able to release green fluorescence protein (GFP) indicating the presence of
intact plasmid DNA. The mechanism of nanoparticle protection of DNA strands
from cleavage is not clear.

2.5 Summary

Photoactive nanomaterials have exhibited different extents of cytotoxic effects. A
summary of the cytotoxicity properties of some typical photoactive nanomaterials
is listed in Table 2.1. In general, the toxic metal-contained nanomaterials, QDs,

Table 2.1 Summary of factors affecting cytotoxicity of photoactive NPs

Nanomaterials	Composition	Size	Surface molecule	Cell line
Cd QDs	Heavy metal, toxic	Increases cytotoxicity as size decreases	Reduces/increases cytotoxicity	Cytotoxicity is related to cell lines
Noble metal NPs				
Ag NPs	Release of ROS, toxic	Decreases cytotoxicity as size decreases	N/A	Cytotoxicity is related to cell lines
Au NPs	Not toxic	N/A	Increase cytotoxicity	Cytotoxicity is related to cell lines
Fluorescent silica NPs	Not toxic	No effect	N/A	No effect

Fig. 2.1 Suggested mechanism of redox active property induced cytotoxicity of heavy metal nanoparticles

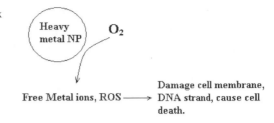

Free Metal ions, ROS ⟶ Damage cell membrane, DNA strand, cause cell death.

exhibited the highest cytotoxic effects among current popular photoactive nanomaterials. For example, the cadmium-contained QDs (CdSe QDs, CdTe QDs) caused acute cytotoxic effects at a low concentration over a short time period ($100\ \mu g \cdot mL^{-1}$ QDs in 24 hr incubation). Redox active metal nanoparticles have also showed significant cytotoxic properties (Fig. 2.1). Silver NPs was an example of such redox metal materials. The fluorescent silica nanoparticles showed the lowest cytotoxic effects among current typical photoactive nanomaterials. At low concentrations, silica nanoparticles showed no apparent cytotoxic effects.

The cytotoxic effects can be reduced by changing nanomaterial compositions and sizes. The sizes of photoactive NPs were in the range of 1–100 nm. The smaller the size the NPs, the lower the cytotoxic effect. The penetration ability of the nanomaterials into cells increased as the size decreased. The larger nanomaterials stayed in the cytosol, while the smaller ones entered the nucleus of cells. Currently three kinds of photoactive nanomaterials, QDs, noble metal NPs, and fluorescent silica NPs, were all able to enter cell membranes but only smaller ones entered the nucleus when observed under fluorescent microscopes. This size effect was observed in each type of nanomaterial. However, silver nanoparticles were reported with a reverse trend.

The coating technique was one effective strategy to reduce or prevent the cytotoxic effects of NPs. Among all the coating materials used, the biocompatible materials, such as silica and BSA, have demonstrated a great potential to reduce the cytotoxicity of nanomaterials. In addition, cytotoxicity of NPs was related to the type of cell. The same type of nanomaterials might exhibit diverse toxic effects to different cell lines.

Different mechanisms of cytotoxic effects of photoactive nanomaterials was reported. The release of toxic cadmium ions was a major hypothesis for the cytotoxic effects of QDs. The presence of redox active elements in metal nanoparticles could cause toxic effects to living cells. Kirchner suggested (Kirchner et al. 2005) that the surface concentration of metal atoms should be used in the study of QDs cytotoxicity instead of using QDs concentration. This surface concentration correlated the ability of QDs to release cadmium ions better than the QD concentration. The smaller QDs had higher surface-to-volume ratios. Comparison of the surface concentrations of differently sized QDs of the same amounts showed, the smaller QDs had a higher amount of surface molecules. Thus, the smaller QDs caused higher cytotoxic effects.

Since the toxicity studies on photoactive nanomaterials described in this chapter has developed rapidly and has attracted great interest in such a short period of time, we expect many novel investigations in this field that will direct developments and applications of photoactive nanomaterials in bioanalysis within the next several years.

Acknowledgements This work was partially supported by Society for Analytical Chemists of Pittsburgh Starter Grant; New Faculty Scholar Awards and Faculty Research Seed Money Awards at the University of North Dakota; and Startup fund from NSF EPSCoR, EPA EPSCoR and BRIN at the University of North Dakota.

References

Braydich-Stolle L, Hussain S, Schlager JJ, Hofmann MC (2005) In vitro cytotoxicity of nanoparticles in Mammalian germline stem cells. Toxicol Sci 88:412–419

Cao YC, Jin R, Mirkin CA (2002) Nanoparticles with Raman spectroscopic fingerprints for DNA and RNA detection. Science 297:1536–1540

Cao YC, Jin R, Nam JM, Thaxton CS, Mirkin CA (2003) Raman dye-labeled nanoparticle probes for proteins. J Am Chem Soc 125:14676–14677

Cao YW, Jin R, Mirkin CA (2001) DNA-modified core-shell Ag/Au nanoparticles. J Am Chem Soc 123:7961–7962

Chang J, Wu H, Chen H, Ling Y, Tan W (2005) Oriented assembly of Au nanorods using biorecognition system. Chem Commun 8:1092–1094

Derfus AM, Chan WCW, Bhatia SN (2004) Probing the cytotoxicity of semiconductor quantum dots. Nano Lett 4:11–18

Dubertret B, Skourides P, Norris DJ, Noireaux V, Brivanlou AH, Libchaber A (2002) In vivo imaging of quantum dots encapsulated in phospholipid micelles. Science 298:1759–1762

Goodman CM, McCusker CD, Yilmaz T, Rotello VM (2004) Toxicity of gold nanoparticles functionalized with cationic and anionic side chains. Bioconjug Chem 15:897–900

Green M, Howman E (2005) Semiconductor quantum dots and free radical induced DNA nicking. Chem Commun: 121–123

He XX, Wang K, Tan W, Liu B, Lin X, He C, Li D, Huang S, Li J (2003) Bioconjugated nanoparticles for DNA protection from cleavage. J Am Chem Soc 125:7168–7169

Hoshino A, Fujioka K, Oku T, Suga M, Sasaki YF, Ohta T, Yasuhara M, Suzuki K, Yamamoto K (2004) Physicochemical properties and cellular toxicity of nanocrystal quantum dots depend on their surface modification. Nano Lett 4:2163–2169

Hussain SM, Hess KL, Gearhart JM, Geiss KT, Schlager JJ (2005) In vitro toxicity of nanoparticles in BRL 3A rat liver cells. Toxicol In Vitro 19:975–983

Ito S, Yoshikawa H, Masuhara H (2001) Optical patterning and photochemical fixation of polymer nanoparticles on glass substrates. Appl Phys Lett 78:4046

Kirchner C, Liedl T, Kudera S, Pellegrino T, Munoz JA, Gaub HE, Stolzle S, Fertig N, Parak WJ (2005) Cytotoxicity of colloidal CdSe and CdSe/ZnS nanoparticles. Nano Lett 5:331–338

Kwon SS, Nam YS, Lee JS, Ku BS, Han SH, Lee JY, Chang IS (2002) Preparation and characterization of coenzyme Q10-loaded PMMA nanoparticles by a new emulsification process based on microfluidization. Colloids and Surfaces, A: Physicochemical and Engineering Aspects 210:95–104

Lagerholm CB, Wang M, Ernst LA, Ly DH, Liu H, Bruchez MP, Waggoner AS (2004) Multicolor coding of cells with cationic peptide coated quantum dots. Nano Lett 4:2019–2022

Li D, He XX, Wang K, He C (2002) Detecting on toxicity of series silica shell inorganic nanoparticles to cells. J Hunan Univ 29:1–6

Liang S, Pierce D, Amiot C, Zhao X (2005) Photoactive nanomaterials for sensing trace analytes in biological samples. Synth React Inorg Met-Org and Nano Met Chem 35:661–668

Lovrić J, Bazzi HS, Cuie Y, Fortin GR, Winnik FM, Maysinger D (2005) Differences in subcellular distribution and toxicity of green and red emitting CdTe quantum dots. J Mol Med 83:377–385

Luo D, Han E, Belcheva N, Saltzman WM (2004) A self-assembled, modular DNA delivery system mediated by silica nanoparticles. J Control Release 95:333–341

Michalet X, Pinaud FF, Bentolila LA, Tsay JM, Doose S, Li JJ, Sundaresan G, Wu AM, Gambhir SS, Weiss S (2005) Quantum dots for live cells, in vivo imaging, and diagnostics. Science 307:538–544

Mirkin CA, Letsinger RL, Mucic RC, Storhoff JJ (1996) A DNA-based method for rationally assembling nanoparticles into macroscopic materials. Nature 382:607–609

Santra S, Yang H, Holloway PH, Stanley JT, Mericle RA (2005) Synthesis of water-dispersible fluorescent, radio-opaque, and paramagnetic CdS:Mn/ZnS quantum dots: a multifunctional probe for bioimaging. J Am Chem Soc 127:1656–1657

Shiohara A, Hoshino A, Hanaki K, Suzuki K, Yamamoto K (2004) On the cyto-toxicity caused by quantum dots. Microbiol Immunol 48:669–675

Wang L, Yang C, Tan W (2005) Dual-luminophore-doped silica nanoparticles for multiplexed signaling. Nano Lett 5:37–43

Yi DK, Selvan ST, Lee SS, Papaefthymiou GC, Kundaliya D, Ying JY (2005) Silica-coated nanocomposites of magnetic nanoparticles and quantum dots. J Am Chem Soc 127:4990–4991

Zhao X, Dytocio RT, Tan W (2003a) Ultrasensitive DNA detection using highly fluorescent bioconjugated nanoparticles. J Am Chem Soc 125:11474–11475

Zhao X, Hilliard LR, Mechery SJ, Wang Y, Bagwe RP, Jin S, Tan W (2004) A rapid bioassay for single bacterial cell quantitation using bioconjugated nanoparticles. Proc Natl Acad Sci USA 101:15027–15032

Zhao X, Tapec-Dytioco R, Wang K, Tan W (2003b) Collection of trace amounts of DNA/mRNA molecules using genomagnetic nanocapturers. Anal Chem 75:3476–3483

Zhou C, Zhao Y, Jao T, Winnik MA, Wu C (2002) Photoinduced aggregation of polymer nanoparticles in a dilute nonaqueous dispersion. J Phys Chem B 106:1889–1897

Chapter 3
Breeching Epithelial Barriers – Physiochemical Factors Impacting Nanomaterial Translocation and Toxicity

Lisa DeLouise, Luke Mortensen and Alison Elder

Abstract With the surging nanotechnology industry, the likelihood of intentional consumer and unintended worker-related skin and lung exposures to various types of nanomaterials is assured. From existing literature, there is clear evidence that some nanomaterials can passively breech epithelial barriers. For skin, mechanical flexing can facilitate penetration of large micron-sized particles and, for both skin and lung, the health status will affect barrier function. Nanoparticle toxicology is, however, an emerging field and inconsistencies in the published literature exist. Inconsistencies should be anticipated as there is currently no standardized set of tests by which nanoparticle toxicity can be determined. Therefore, the question of nanomaterial toxicity resulting from unintended epithelial permeation remains open. In vitro cytotoxicity studies clearly indicate that nanomaterials are toxic to skin and lung cells under certain conditions. The relevance of these results is difficult to extrapolate, as there is a presumption of epithelial permeation. This chapter discusses what is known and what is not known about physiochemical factors impacting nanomaterial translocation and toxicity.

Contents

L. DeLouise (✉)
Departments of Dermatology, Biomedical Engineering and Environmental Medicine, University of Rochester Medical Center, Rochester, NY, USA
e-mail: Lisa_DeLouise@urmc.rochester.edu

T.J. Webster (ed.), *Safety of Nanoparticles*, Nanostructure Science and Technology, DOI 10.1007/978-0-387-78608-7_3, © Springer Science+Business Media, LLC 2009

3.1 Introduction

Although many definitions exist for nanomaterials, a very basic one describes an object with at least one dimension having a diameter less than 100 nm. This definition can obviously apply to spherical particles as well as fibrous materials. Nanomaterials are also those that are created via "the manipulation of materials at the atomic, molecular, and macromolecular scales, where the properties differ significantly from those at a larger scale" (The Royal Society and Royal Academy of Engineering, 2004; Oberdörster et al. 2005; Nel et al. 2006; Klein 2007). A characteristic of nanomaterials is that their surface area per unit mass increases as size decreases, as does the percentage of atoms that can be found at the surface of the material. This may partly explain why unique properties can be found at the nanoscale that are not present in the same bulk material, such as the size dependent fluorescence emission frequency of semiconductor quantum dot (QD) nanomaterials.

Advances made in the synthesis and control of engineered nanomaterial properties have expanded their use in numerous diverse applications ranging from medicine (diagnostic imaging, targeted drug delivery, biosensors) to energy (solar cells, catalysts) and consumer goods (cosmetics, inks, electronics). A consequence of this nanotechnology boom is an increasing risk of unintended consumer and occupational exposure to nanomaterials. Two likely routes of exposure are contact with skin and the respiratory tract. Unfortunately, current understanding of nanomaterial penetration through epithelial tissue and their potential toxicity is poor (Curtis et al. 2006; Hardman 2006). Generating conclusive data are a daunting challenge that is confounded by the compositional diversity of nanomaterials and the wide range of synthetic processes used to create them (Pulskamp et al. 2007). Nonetheless, trends are emerging in recent literature that suggest that correlations exist between tissue penetration and key physiochemical characteristics. Translocation and toxicity depend on specific cellular interactions and intrinsic nanomaterial factors, such as biochemical stability and their potential to generate reactive oxygen species (ROS) (Nel et al. 2006; Cedervall et al. 2007). These trends are reviewed in this chapter. We begin Section 3.2 with an introduction to important physiochemical properties that are likely to affect nanomaterial interactions with epithelia. This is followed by a discussion of the current understanding of nanomaterial interactions with tissues of the skin in Section 3.3 and the respiratory tract in Section 3.4. Both sections first introduce the essential microanatomical and cellular features that are critical for understanding how nanomaterials can breech these epithelial barriers. This is followed by a discussion of key themes emerging from recent literature on nanomaterial penetration, translocation, and toxicity.

3.2 Nanomaterial Physiochemical Properties

The physiochemical properties of nanomaterials that have emerged as important determinants of uptake and toxicity include size, surface charge, surface energy (hydrophobicity/hydrophilicity), and chemical composition. Of these, particle size is of utmost importance in assessing epithelial penetration and toxicity.

3.2.1 Nanomaterial Size

Epithelial barriers are formed in part by tight intercellular junctions that have aqueous spaces through which penetration via diffusion processes can occur (Anderson 2001). However, the aqueous spaces are small (~4 nm diameter) and penetration of larger particles is physically hindered unless junctions are leaky due to damage, disease, or inflammation. In fact, the latter may be induced upon nanomaterial contact. Once particles enter tissues, effective clearance mechanisms also depend on size. For example, nanomaterials exceeding ~4–6 nm diameter that gain access to the circulation by breeching epithelial barriers or by intravenous (Choi et al. 2007; Fischer et al. 2006) injection are considered too large to be excreted in urine (Smith et al. 2004). Recent studies find that after 12 hrs, only 0.25% of the QDs (~37 nm) injected intradermally accumulated in kidney tissue (Gopee et al. 2007); however, detection of QDs in urine was not examined. A higher percentage of circulating nanomaterials are cleared by uptake into phagocytic cells in the organs of the reticuloendothelial system (RES), most notably the liver and spleen (Fischer et al. 2006; Gopee et al. 2007). However, the clearance rate by the RES system also depends on size (Lutz et al. 1989). Studies show that larger particles (radius 250 nm) are phagocytosed faster than smaller particles (radius 25 nm) (Holmberg et al. 1990).

3.2.2 Nanomaterial Composition

Nanomaterial composition (surface coating and core) is an important determinant of toxicity in that it influences cellular uptake, RES clearance rate, and biochemical compatibility. Nanomaterials that come into contact with epithelial tissues may be taken up by cells via receptor-mediated endocytosis or nonspecific pinocytosis. Hydrophilic coatings like polyethylene glycol (PEG) and BSA tend to delay uptake (Åkerman et al. 2002; Zahr et al. 2006). Once inside a cell they may remain sequestered or they may be degraded depending upon their biochemical stability. A growing body of literature suggests that endocytic uptake and subsequent intracellular trafficking into acidic vesicles is an important mechanism for potentiating cytotoxicity, particularly when the nanomaterial surface coatings and/or core composition are pH sensitive (Chang et al. 2006; Lovrić et al. 2005a, b; Ryman-Rasmussen et al. 2007b). Also the intrinsic ability of nanomaterials to generate ROS and to cause oxidative stress has been linked to greater cytotoxicity (Sayes et al. 2005; Ipe et al. 2005; Lovrić et al. 2005a; Tsay and Michalet 2005; Nel et al. 2006). Composition also influences cytotoxicity by impacting ROS generation, surface charge, and surface energy (Derfus et al. 2004; Sayes et al. 2004; Gupta and Gupta 2005; Hoshino et al. 2004), as discussed below.

3.2.3 Nanomaterial Surface Charge

In a biological environment, surface charge will influence nonspecific adhesion of proteins on the nanomaterial surface (Choi et al. 2007), its permeation through tight junctions, and cellular uptake. Cell membranes are typically negatively charged, although positive domains exist (Ghitescu and Fixman 1984). At physiological pH ($\sim$7.3), aqueous pore channels found in tight junctions are lined with anionic charges, a condition that slightly favors permeation of cations (Anderson 2001). Recently, it was shown that positively charged polymer nanomaterials ($\sim$200 nm diameter) were phagocytosed by mouse macrophages to a greater extent than negatively charged particles (Zahr et al. 2006). Studies of cell adhesion to charged polymeric microcapsules (5 μm diameter.) showed that positively charged microcapsules exhibited greater de-adhesion forces ($\sim$25 pN) compared to negative microcapsules ($\sim$18 pN) (Javier et al. 2006). Positively charged microcapsules were also nonspecifically taken up by cells at a higher rate than negatively charged ones (Javier et al. 2006). Contrary to expectations based on electrostatic arguments, others have found that phagocytosis of hydrophilic polystyrene particles ($\sim$1 μm diameter.) by mouse macrophages increased with negative charge (Gbadamosi et al. 2002). This result is supported indirectly by a study that found anionic particles induce a greater inflammatory response (Cui and Mumper 2001) which activates phagocytic cells (Trinchieri et al. 1993). These observations illustrate the difficulty in applying generalized concepts for predicting nanomaterial permeation, cell uptake, and toxicity in biological systems.

3.2.4 Nanomaterial Surface Energy

Lastly, surface energy has been found to greatly impact how nanomaterials interact with biomolecules and tissue (Cedervall et al. 2007). In an aqueous environment low surface energy coatings (hydrophobic) are particularly prone to nonspecific adsorption as proteins denature to expose their hydrophobic core. They also show surfactant-like properties, which may help extract or disorganize lipid components of cell membranes, lowering barrier function enabling enhanced epithelial penetration (Baroli et al. 2007). High surface energy coatings (hydrophilic), particularly those that produce a weakly negative or neutral surface charge, are ideal for resisting protein adsorption and cell uptake (Zahr et al. 2006; Derfus et al. 2004). It is important to note that nonspecific biomolecule adsorption is a competitive and dynamic process. Known as the Vroman effect and initially described for blood plasma proteins (Turbill et al. 1996), nonspecific binding begins with the adsorption of abundant low molecular weight proteins (e.g., albumin) that diffuse quicker to the material surface. In time, high molecular weight proteins (e.g., fibrogenin) of lower concentration accumulate on the surface as they are harder to displace. As such, nanomaterials engineered with noncovalent coatings to affect a specific function (e.g., protect against degradation, target delivery, resist cellular uptake, or clearance, etc.) may in fact change with time in vivo depending upon tissue location (Klein 2007; Cedervall et al. 2007).

3.3 Skin

3.3.1 Microanatomy and Mechanisms of Particle Translocation

The skin is the largest organ in mammals. It provides many functions, the most important of which is protection from the environment. Skin is stratified squamous epithelial tissue. It has evolved a multilayer physical architecture grossly comprised the innermost subcutis, the dermis, and the outermost epidermis covered by the stratum corneum layer (Fig. 3.1). Each layer consists of different cell types, biomolecules, and skin appendages (follicles and glands) that uniquely work together to maintain the barrier function. Key features of this architecture relevant to nanomaterial penetration are detailed in Fig. 3.1.

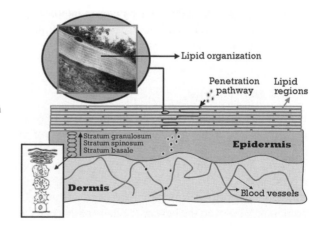

Fig. 3.1 Schematic illustrating the multilayer architecture of skin. The layers of the stratum corneum provide a strong barrier to penetration. The flow of terminally differentiating keratinocytes from basale to the stratum corneum adds to the barrier function (Adapted from Bouwstra and Ponec 2006)

3.3.1.1 Physical Architecture

The subcutis and the dermis are the innermost portions of the skin. They both provide nutrient and waste transport for the epidermis along with mechanical protection and thermal insulation. The dermis shares structural features with the subcutis, such as the presence of adipose tissue, collagen–glycosaminoglycan complexes, loose connective tissue, and elastic proteins. A variety of skin appendages reside in the dermis including hair follicles, nerve endings, and secretory glands (sweat, sebaceous). Secretory glands provide a supporting role in preventing skin penetration. Sebaceous glands release lipids that keep the skin flexible and provide a barrier to hydrophilic substances. Sweat glands release salty and slightly acidic (pH ~5.0) secretions which can agglomerate particles, thus help to prevent penetration.

The basement membrane separates the dermis and the epidermis. It consists of closely packed collagen, laminin, fibronectin, and other cell adhesion molecules. The basement membrane is bordered on the apical side by the epidermis, which consists of several layers including the stratum basale, stratum spinosum, stratum granulosum, and the stratum corneum which is the main permeation barrier. The stratum basale is made up of keratinocytes (~90%), which are responsible for maintaining the proliferative potential of the skin. Keratinocytes divide in the stratum basale and migrate through the upper layers undergoing terminal differentiation. The outward cellular movement contributes to the barrier function of skin by pushing substances that have breeched the stratum corneum back out towards the skin surface. As keratinocytes migrate to form the stratum corneum, they become corneocytes. They lose their nuclei, become flattened, and produce lamellar granules containing lipids and protein granules containing keratin, filaggrin, and loricrin. These proteins are anionically charged and are key components in forming a strong mechanical barrier. Skin surface charge provides electrostatic repulsion of anionic penetrants (Kim et al. 2006; Wagner et al. 2003; Baroli et al. 2007). Transmembrane proteins (E-cadherins and desmogleins) link corneocytes forming tight intercellular junctions that are internally linked to intermediate filaments, keratin, and actin. These intercellular connections provide physical strength and unity of the stratum corneum barrier. The final result is a brick and mortar structure (Fig. 3.1) consisting of 12–16 layers of corneocytes surrounded by exocytosed intercellular lipid lamellae composed of ceramides, fatty acids, and other lipids. The stratum corneum layer provides the principal barrier function of skin.

3.3.1.2 Hair Follicles

Hair follicles occur throughout the skin. They play an important role in mitigating permeation, as their physical invaginated structure provides a niche for mechanical accumulation and storage of substances (see "Hair follicles are Efficient Reservoirs for Accumulation of Nanomaterials"). Each follicle is comprised of an inner root sheath (IRS), outer root sheath (ORS), and a fibrous sheath (FS) (Fig. 3.2). These provide a barrier function; however, it is not as fully developed as the stratum corneum. Moreover, the base of the hair follicle is located in the dermis and is fed by

Fig. 3.2 The hair follicle has a modified barrier structure that leaves a greater potential for penetration of chemical substances (Toll et al. 2004)

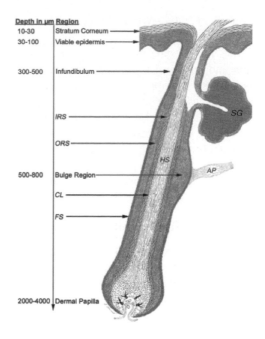

the bloodstream and lymph. This architecture makes hair follicles a potential portal for nanomaterial penetration with potential for systemic access (Toll et al. 2004; Meidan et al. 2006). In fact, studies comparing the efficacy of transdermal drug delivery through normal and scarred tissue (hair follicles and glands do not regenerate in deep tissue injury) confirm the importance of follicles in skin permeation (Hueber et al. 1992; Illel et al. 1997).

3.3.1.3 Skin Immune Cells and Injury Response

Despite its strength and resistance to most substances, the stratum corneum can be disrupted by a variety of conditions. Skin exposure to chemicals, organic solvents, detergents, tape stripping, UV radiation, mechanical injury, and disease all affect stratum corneum barrier function. In humans, the inflammatory response to mild injury (excluding disease) follows a similar healing course. Cytokine production induces a burst of cellular activity that activates professional immune cells (macrophages and Langerhans cells) residing in the epidermis and dermis. These cells work to provide a secondary blockade protecting against systemic permeation of substances. Macrophages phagocytose foreign substances, microbial invaders, and particles as large as 0.5–3.5μm in diameter (Tinkle et al. 2003), degrading them in lysosomes. Langerhans cells are professional antigen presenting cells. They phagocytize foreign substances, move them to lymph nodes, and present antigen to T-cells (Proksch and Brasch 1997). They are highly concentrated in skin around hair follicles, allowing them to be targeted for vaccines and drug delivery

(Vogt et al. 2006). Within hours of injury, E-cadherins are rapidly down-regulated, disrupting cell tight junctions to enable increased cell motility (Brouxhon et al. 2007). At the same time, upregulation of COX-2 induces keratinocyte proliferation and differentiation. New corneocytes form and over the course of 2–3 days, skin thickens (Tripp et al. 2003; Jiang et al. 2007). The stratum granulosum quickly fills in holes with "mortar" by increasing production of cholesterols and fatty acids and secreting lamellar bodies in its upper levels (Menon et al. 1992). Skin barrier function is usually restored to 50% within 12 hrs and full function within 3–4 days (Ghadially et al. 1995; Jiang et al. 2007). This response returns homeostasis and skin barrier function rapidly; however, during the healing process, down regulation of E-cadherins can cause a marked increase in permeability.

3.3.1.4 Summary of Skin Permeability and the Polar and Lipid Pathways

The main factors affecting skin permeability are the characteristics of the substance in contact with the skin and the health status of the skin barrier. The primary determinants of permeability are particle size and surface energy (hydrophobicity/hydrophilicity). The importance of particle surface charge in skin permeation has also been established in studies of transdermal drug delivery (Parisel et al. 2003; Lee et al. 2007). Surface energy generally determines whether permeation occurs via the polar or lipid pathway. Polar substances move para- and trans-cellularly in association with proteins through tightly linked corneocytes or para-cellularly between corneocytes through polar channels formed in lipid lamellae (Dayan 2005). Both mechanisms are important however, finite element modeling and literature data suggest that permeation primarily occurs transcellularly through corneocytes, causing very high lag times for polar substances (Barbero and Frasch 2006). Skin tends to be far more permeable to hydrophobic substances, whose main route of entry is intercellularly between corneocytes through the lipid lamellae (Fig. 3.1). Accumulation and permeation via the hair follicle are also important processes for particulate materials as is discussed in "Hair follicles are Efficient Reservoirs for Accumulation of Nanomaterials".

3.3.2 Lessons from the Study of Ultrafine Particle Interactions with Skin

In recent years, studies of nanomaterial–skin interactions have escalated owing to their increasing use in technology and commercial products. For example, the antimicrobial properties of silver nanomaterials are exploited in wound care products (bandages, masks), food containers, and refrigerators (Morones et al. 2005). Magnetic and metal oxide nanomaterials are used in UV sun screens and cosmetic products (Sincai et al. 2007; Salleh 2004; Cross et al. 2007). Interestingly, metal oxide containing sunscreens are frequently prescribed to patients to protect photosensitized skin following treatment for actinic keratosis and skin cancer, despite treatment-induced defects in skin barrier function (Schwarz et al. 2001; Fien and Oseroff 2007). The bulk of our current understanding of nanomaterial sensitization and penetration through skin comes from in vitro cell culture and ex vivo human and pig skin studies. Although these models have their strengths, comparatively few in vivo studies employing human or animal skin exist. In the following, we provide key lessons gleaned from recent research.

3.3.2.1 In vitro Skin Cell Studies

In vitro studies are ideally suited for quantifying nanomaterial cytotoxicity and investigating pharmacokinetic mechanisms of cell uptake and death (e.g., apoptosis vs. necrosis). Skin cell studies (keratinocytes and fibroblasts) have been conducted with a variety of nanomaterial types including carbon-based fullerenes and nanotubes (Shvedova et al. 2003; Sayes et al. 2004, 2005, 2006a; Ding et al. 2005; Monteiro-Riviere et al. 2005a, b; Tian et al. 2006; Witzmann et al. 2006), QDs (Zhang et al. 2006; Ryman-Rasmussen et al. 2007a, b), polymers (Weyenberg et al. 2007), metals (Berry et al. 2003, 2004; Gupta and Gupta 2005; Lam et al. 2004; Auffan et al. 2006), and metal oxides (Sayes et al. 2006b). For example, in vitro studies of silver nanoparticles were reported to greatly reduce the viability of cultured keratinocytes (Lam et al. 2004). Likewise, citrate-coated gold nanoparticles and magnetic nanomaterials caused cytotoxicity in human dermal fibroblasts (Pernodet et al. 2006; Gupta and Gupta 2005). Cytotoxicity of titanium oxide nanoparticles towards human dermal fibroblasts correlated with particle phase composition (anatase > rutile) and ROS generating capacity (Sayes et al. 2006b). These results follow trends consistent with other cell types that link oxidative stress (Nel et al. 2006; Tsay and Michalet 2005) and proinflammatory cytokine generation (Ryman-Rasmussen et al. 2007a) to cytotoxicity. They also provide important insight for the design of coatings that can prevent nanomaterial degradation (Derfus et al. 2004; Chang et al. 2005), cell uptake (Ryman-Rasmussen et al. 2007b) and, hence, cytotoxicity (Sayes et al. 2004).

3.3.2.2 Ex vivo and In vivo Skin Studies

A caveat of in vitro studies is that they are typically done in a 24–48 period at dose levels that far exceed what might be expected to occur from an incidental exposure. They also presume that nanomaterials breech the stratum corneum barrier and penetrate to interact with keratinocytes and fibroblasts, the two most common cell types found in living epidermis and dermis, respectively. To ascertain the relevance of the in vitro skin cell studies, a definitive answer to the question of skin permeability to nanomaterials must be sought. Ex vivo and in vivo experiments conducted with skin have yielded conflicting results. In the following, we discuss the current understanding of whether nanomaterials can breech the stratum corneum barrier and provide rationale for the observed discrepancies.

Ex vivo studies typically employ human skin resected from breast reduction or abdominoplasty surgeries. Alternatively, pig skin is used as a good human skin model exhibiting similar permeability (Dick and Scott 1992; Gamer et al. 2006) and follicular structure (Lademann et al. 2007). Permeation through ex vivo skin has traditionally been investigated using a Franz diffusion cell (Kohli and Aplar 2004) and/or tape stripping combined with immunohistology (Lademann et al. 2006b) employing a variety of spectroscopies. Franz diffusion studies involve clamping a skin sample between donor and receptor compartments of a vertical diffusion cell. The tissue is perfused from below in a buffer solution at 37°C with the test formulation applied to air-exposed stratum corneum. The electrical resistivity or impedance of the skin sample is often measured before analysis to ensure results are not confounded by tissue defects from vehicle artifacts or subcutaneous fat removal (Baroli et al. 2007; Cross et al. 2007). Tape stripping is a process used to evaluate the distribution of topically applied substances in the stratum corneum. An adhesive strip is pressed and removed successively from the same area of treated skin. This process removes corneocytes and the substances that were applied in a layer by layer fashion. The composition and spatial distribution of substances adhered to the tape can then be determined (Lademann et al. 2007). In recent years, confocal microscopy has been increasingly used for analysis of whole skin (Tinkle et al. 2003; Rouse et al. 2007).

Because of the importance of developing topical drug and virus delivery systems (Destree et al. 2007; Liu et al. 2006; Lademann et al. 2006a; Vogt et al. 2006), considerably more research exists on the interaction of polymers and lipid nanomaterials with skin. However, with increasing toxicity concerns, the investigation of other types of nanomaterials (metals, QDs, carbon-based nanomaterials) is rising. From this body of literature, several important themes are emerging that appear to be independent of particle composition.

Hair Follicles are Efficient Reservoirs for Accumulation of Nanomaterials

Hydrophilic carboxy terminated polystyrene fluorescent beads (20 nm diameter) applied ex vivo to porcine skin showed superior follicular deposition compared to larger beads (200 nm diameter) (Alvarez-Roma et al. 2004). No evidence for the penetration into epidermis was noted; however, particles were exposed to skin for a maximum time of only 2 hrs before analysis. More recently, the penetration of rigid iron oxide core nanomaterials ($\sim$5–23 nm) through ex vivo human skin was investigated (Baroli et al. 2007). Franz diffusion cell measurements did not detect permeation through the thick skin sample at time points up to 24 hrs. However, EDS-SEM analysis confirmed the presence of nanomaterials throughout the stratum corneum and occasionally in the uppermost strata of the viable epidermis extending to 30–100 μm deep. Normal skin invaginations and hair follicles were observed to be privileged locations for metal nanomaterial accumulation. These observations are corroborated by results from an in vivo study investigating the protective effect of magnetic nanomaterial fluids ($\sim$0.03 mg/cm^2) on prolonged UV radiation on mouse ears (Sincai et al. 2007). Results showed that nanomaterials penetrate into the dermis via hair follicles (IRS and cuticular layer (CL) Fig. 3.2). Cytohistological and TEM results at higher skin surface concentrations ($\sim$0.1 mg/cm^2) indicate that dermal fibroblasts and macrophages were loaded with magnetic nanomaterials.

Hair Follicle Penetration Depth is Size Dependent

Studies show that the accumulation and penetration into hair follicles is size dependent. For example, carboxylate latex particles of different size (40, 750, and 1500 nm) were applied to ex vivo human skin following tape striping to open hair follicle infundibulum to enhance uptake. After 16 hrs, only the 40 nm polymer particles were able to passively penetrate through follicular epithelium to the perifollicular dermis of the hair shaft (Fig. 3.3). Some reached the sebaceous duct $\sim$225 μm into the dermis. Flow cytometry proved that Langerhans cells (Cd1a+) isolated from tissue had phagocytosed the 40 nm particles but not the larger particles. Interestingly, Langerhans cells cultured in vitro internalized all particle sizes (40,

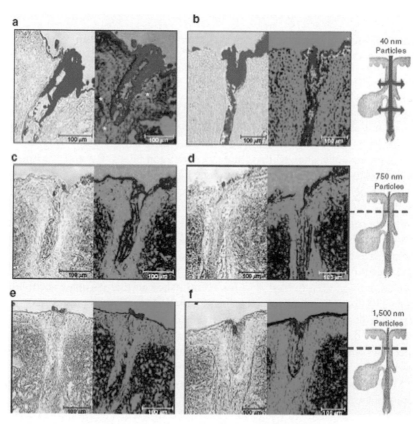

Fig. 3.3 Laser scan microscopy of cryosectioned skin samples illustrating that 40 nm (**a,b**) but not 750 nm (**c,d**) or 1500 nm (**e,f**) fluorescent polymer particles (*red*) penetrate into tissue surrounding hair follicles indicated by white * in (**a**) (Adapted from Vogt et al. 2006)

750, and 1500 nm), but only the smaller particles (40 nm) were uptaken in ex vivo skin. This study provides definative evidence that polymer nanomaterials can breech an intake stratum corneum barrier of ex vivo skin in a size-dependent manner.

Accumulating nanomaterials into hair follicles may provide a functional benefit for drug delivery. Recent studies of transdermal drug release from 40 to 130 nm diameter polymer particles through ex vivo guinea pig skin (Shim et al. 2004) showed that smaller particles were more effective at delivering drugs. This was attributed to enhanced accumulation and deeper penetration of the smaller particles into hair follicles. Greater accumulation and large surface area enhanced release. Moreover, particles that accumulate in hair follicles are retained longer in skin. When fluorescein conjugated hydrogel polymer particles (320 nm) accumulated in the hair follicles of ex vivo porcine and in vivo human skin samples, they were efficiently stored there for up to 10 days, whereas fluorescein dye applied in pure form was eliminated after 4 days (Lademann et al. 2007). Textile contact and desquamation remove particles from the skin surface, whereas sebum production can promote release from follicles.

Mechanical Flexing Enhances Follicular Accumulation and Stratum Corneum Penetration

Several studies have established that mechanical force enhances the accumulation of nanomaterials (Lademann et al. 1999; Rouse et al. 2007) and micron-sized particles (Tan et al. 1996; Tinkle et al. 2003; Toll et al. 2004; Lademann et al. 2007) into hair follicles and penetration through the stratum corneum. A seminal study performed by Tinkle et al. (2003) on micron-sized particles employing scanning confocal microscopy, showed that fluorescent dextran beads (0.5 and 1.0 μm) penetrated into the epidermis of ex vivo human skin in 56% of the samples (11 tested) and into the dermis of 18% after 60 min of flexing. They noted, however, that only a small percentage of the 0.5 and 1.0 μm fluorospheres applied actually penetrated and no evidence that the larger 2 and 4 μm particles penetrated the stratum corneum was found. This further supports the notion that penetration and follicular accumulation are size dependent, even with mechanical flexing. Interestingly, dye-labeled carboxy-terminated polystyrene microspheres (0.75–6 μm) topically applied and massaged onto ex vivo human skin for 10 min showed that particles accumulated deep (~2300 μm) into hair follicles, but penetration into epidermal layers was not observed (Toll et al. 2004). These two studies confirm the penetration of polymer beads is size dependent, but the inconsistencies in the cut-off size result from differences in skin integrity, the mechanical flexing technique and/or the particle skin exposure time. Neither study quantified the surface properties of the polymer particles. More recently, the effect of flexing on the penetration of peptide functionalized fullerene nanomaterials (3.5 nm core diameter) through ex vivo pig skin (400 μm thick) was investigated (Rouse et al. 2007). The fullerenes penetrated extensively through flexed skin deep into the dermis in just 1.5 hrs. Passive penetration through unflexed skin was also observed, but after 8 hrs, fullerene nanomaterials were only detected in the epidermis. TEM studies provide evidence that fullerene migration through the skin occurs intercellularly, utilizing the lipid pathway. This is consistent with the intrinsic hydrophobicity of fullerenes despite the attached amino acid.

Skin Permeation Depends on Surface Charge

The issue of nanomaterial surface charge on skin penetration was specifically addressed in a study of positive, negative, and neutral fluorescent latex particles (50, 100, 200, 500 nm) passively applied to ex vivo pig skin mounted in a Franz diffusion cell (Kohli and Aplar 2004). Fluorescence microscopy showed that most particles remained on the skin surface. After 4 hrs only a small fraction (~0.15%) of the negatively charged 50 and 500 nm particles were detected in the perfusate, indicating they had penetrated through the entire skin sample. This result contradicts the permselectivity skin model that predicts that cationic particles are more favored to permeate skin (Marro et al. 2001) because of electrostatic attraction to cell membranes (Javier et al. 2006) and the acidy of the stratum corneum (Wagner et al. 2003). The authors argue that permeation through stratum corneum results from repulsive interactions in the lipid lamellae. Only the negatively charged 50 and 500 nm particles exhibit sufficient surface charge density to induce the requisite repulsive force. While this argument is plausible, skin inflammatory factors may also have influenced their result. It is known that topically applied cationic nanomaterials can induce inflammation (Badea et al. 2007); however, negative particles were shown to induce a higher immune response (Cui and Mumper 2001). Skin inflammation causes a rapid down regulation of E-cadherins, which are essential proteins in forming tight junctions between corneocytes (Brouxhon et al. 2007). Topical application of substances that induce skin inflammation can compromise barrier functions enabling greater penetration. Again, this study points to the difficulties in applying generalized concepts to complex biological systems.

Nanomaterials Generate Proinflammatory Cytokines in Skin

The interaction of copper, nickel (Hostynek and Maibach 2004; Hostynek et al. 2001), and carbon-based materials (Eedy 1996; Shvedova et al. 2003) with skin has long been a concern for allergic contact dermatitis. Indirect evidence that topically applied nanomaterials can activate the NF-kβ pathway (Manna et al. 2005) causing proinflammatory cytokine release in skin comes from several in vitro studies of keratinocytes exposed to fullerenes (Rouse et al. 2006), carbon nanotubes (Monterio-Riviere et al. 2005a; Ding et al. 2005; Witzmann et al. 2006; Smart et al. 2006), and semiconductor QDs (Ryman-Rasmussen et al. 2007a). Studies show that both cytokine production and permeation of QDs through skin depend on QD size and surface chemistry (COOH, PEG-NH$_2$, and PEG terminations) (Ryman-Rasmussen et al. 2006, 2007a). Extensive permeation of QDs (Ryman-Rasmussen et al. 2006) and fullerenes (Rouse et al. 2007) is observed exhibiting a diffusive penetration through the epidermis and into the dermis of ex vivo pig skin. However, a recent in vivo study of COOH-terminated QD topically applied via glycerol to SKH-1 hairless mice skin produced contrasting results (Mortensen et al. 2007). Extensive

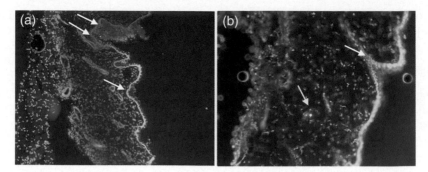

Fig. 3.4 In vivo study of quantum dot penetration through mouse skin. Cryosection with DAPI stain illustrating (**a**) accumulation of quantum dots in hair follicles and on the stratum corneum (*white arrows*) and (**b**) penetration of quantum dots beyond the stratum corneum (*yellow arrow*) (Mortensen et al. 2007).

follicular accumulation but limited penetration beyond the stratum corneum was observed (Fig. 3.4) even when UV radiation exposure was employed to induce a skin inflammatory response which is known to induce changes in skin morphology and barrier function (Pentland et al. 1999; Thomas-Ahner et al. 2007; Brouxhon et al. 2007). The in vivo results corroborate the limited permeation observed of similarly sized metal and polymer nanomaterials (Sincai et al. 2007; Baroli et al. 2007; Tinkle et al. 2003; Kohli and Aplar 2004; Vogt et al. 2006). Whether the extensive permeation through ex vivo pig skin is exacerbated by an inflammatory response or inadvertent factors such as degradation of barrier function through use of a basic vehicle (pH 8–9) (Baroli et al. 2007) remains to be fully understood.

Do Metal Oxide Particles Breech the Stratum Corncum?

While the preponderance of existing data suggests that fullerene, QD, metal, polymer nanomaterials, and even micron-sized polymer particles aided by mechanical force can breech the stratum corneum and penetrate into the epidermis and dermis (albeit a small fraction of what is applied), existing ex vivo and in vivo literature on rigid metal oxide nanomaterials (TiO_2 and ZnO) indicate that they do not penetrate through skin. This question is significant because of the use of metal oxide nanomaterials in consumer UV sunscreens and daily use cosmetic products (Gamer et al. 2006; Cross et al. 2007; Salleh 2004).

Metal oxide particles used in cosmetic formulations span a range of sizes, shapes, and surface coatings. ZnO and TiO_2 are typically formulated in skin lotions as oil/water emulsions. Process technologies produce nanometer size particles, but once they are formulated into a lotion, agglomeration, and aggregation may occur (Fig. 3.5). Coatings and surfactants are used to aid in dispersing metal oxides; however, particle sizing in the formulation remains difficult to characterize using standard methods (Cross et al. 2007). This is particularly true after application to skin

Fig. 3.5 TEM image illustrating agglomeration of zinc oxide nanomaterials in water phase of cosmetic oil/water formulation that arrange around oil droplets (Adapted from Gamer et al. 2006)

where surface pH, salts, and oil may affect their dispersion and size. These issues are rarely considered in the literature.

One of the first studies to investigate TiO_2 skin penetration was performed on human patients scheduled to have skin reduction surgery (Tan et al. 1996). Sunscreen was applied 2–6 weeks prior to the operation. After excision, the stratum corneum was removed by tape stripping and titanium in the epidermis and dermis was quantified using mass spectrometry. Results showed more titanium in the epidermis and dermis of subjects treated with sunscreen relative to controls. While initially interpreted as evidence for penetration, the result was later shown by another group (Lademann et al. 1999) to be consistent with particles trapped in hair follicles. Tape stripping was used to quantify the distribution of TiO_2 (17 nm) formulated in an oil-in-water emulsion applied to human skin multiple times a day for 4 days (Lademann et al. 1999). Results show that metal oxide nanomaterials were localized only on corneocytes in the upper layers of the stratum corneum and accumulated into hair follicles. In more recent work, the penetration through ex vivo porcine skin of ZnO (80 nm) and TiO_2 needle like particles (30–60 nm × 100 nm) formulated in actual sunscreen lotion was investigated using Franz diffusion and tape striping (Gamer et al. 2006). Almost total recovery of the amount of zinc and titanium applied was achieved after a 24 hr application, indicating no penetration. A similar conclusion was reached by Cross et al. (2007) in a study of ZnO particles (25–30 nm) applied to ex vivo human skin. Particles were not detected in lower stratum corneum layers or the viable epidermis by electron microscopy after a 24 hr application. Finally, optical and electron microscopy methods were used to investigate the in vivo penetration of different TiO_2 (hydrophobic and hydrophilic) formulations applied to the forearm of human subjects for 6 hrs (Schulz et al. 2002). Analysis of punch biopsy samples showed no evidence of penetration or dependence on particle size, shape, or surface energy. Rather, TEM analysis showed the formation of a continuous film of agglomerated particles on the outermost layer on the skin surface.

Collectively these studies indicate that metal oxides do not penetrate skin. This raises the question as to why this is so, considering most other nanoparticle types (polymers, QDs, metals and carbon nanotubes) have been found to permeate skin. Is this a fortuitous result of particle agglomeration combined with their intrinsic hydrophobicity? Do metal oxide particles partition into the stratum corneum, where they become trapped in the lipid lamella and remain until desquamation or sebaceous secretions remove them from follicles? Efficient transdermal drug delivery has been correlated with skin hydration (Wissing and Muller 2002). This indirectly suggests the importance of accessing the polar pathway in penetrating the skin barrier which would hinder hydrophobic particle penetration. Are there physiochemical factors other than size, surface charge, and surface energy that are important in considering nanomaterial penetration? (For example, the nanoparticle mechanical properties.) It has been recently found in studies measuring the distribution profile of elastic and rigid vesicles (~115 nm diameter) in human skin that elastic particles penetrate deeper under identical conditions (Honeywell-Nguyen et al. 2004). Must particles be compliant to squeeze through the stratum corneum? Are metal oxide nanoparticles less compliant than metals of semiconductor QDs? Alternatively, are metal oxide nanomaterials simply too difficult to detect in skin (Baroli et al. 2007; Schulz et al. 2002), leading to incorrect conclusions regarding penetration? Clearly more research is needed to resolve this discrepancy and to gain a quantitative understanding of the extent and mechanisms of nanoparticle skin penetration.

3.4 Respiratory Tract

3.4.1 Microanatomy and Mechanisms of Nanomaterial Deposition in the Lung

A second likely route for unintended nanomaterial exposure is the respiratory tract. To understand how inhaled nanomaterials might interact with lung cells, it is critical to understand where and how they deposit in the respiratory tract. Along with this information, some knowledge of the cell types present and the nature of the lung lining fluid in a given region is useful. The International Commission on Radiological Protection developed a model (ICRP 1994; Bailey 1994) that predicts a high fractional deposition efficiency for aerosolized nanomaterials in the alveolar regions of the lung; however, significant amounts also deposit in the nasopharyngeal–laryngeal and tracheobronchial regions (Fig. 3.6). In general, the cell type distribution of the respiratory tract epithelium becomes less complex and the cells "flatten out" from the conducting airways to the alveoli. For example, ciliated pseudostratified columnar epithelial cells and mucous-producing goblet cells line the conducting airways down to small bronchi, whereas the epithelium in the alveoli is one squamous cell thick in most places (Junquiera et al. 1992; Ross and Romrell 1989).

The nose filters very large and nanosized particles and it absorbs gases. Of particular note in the olfactory epithelium of the nose is the presence of ciliated

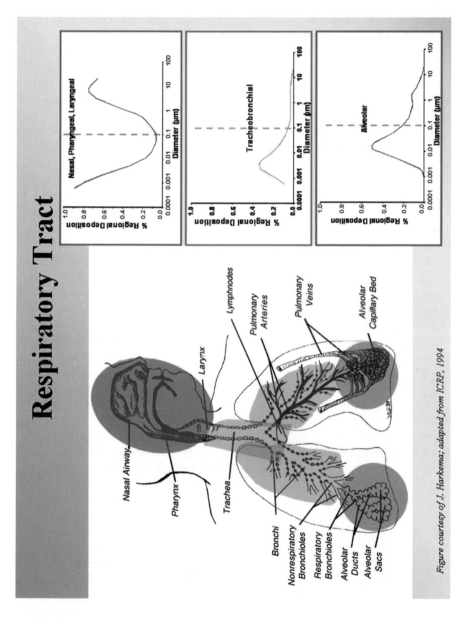

Fig. 3.6 Pathways of particle deposition in the respiratory tract (from Oberdörster et al. 2005)

olfactory cells, which are bipolar neurons that are continuous with the olfactory bulb in the central nervous system. These cells are of interest as a potential portal for nanomaterials to enter the brain. Solid particle transport along the olfactory nerve into the olfactory bulb has been demonstrated via electron microscopy with intranasally applied polio virus and Ag-coated colloidal (50 nm) Au (Bodian and Howe 1941a, b; DeLorenzo 1970). Strong evidence for this same process based on kinetics and solubility characteristics (at neutral pH) has also been shown using chemical means for laboratory-generated Mn oxide and ^{13}C ultrafine particles (Elder et al. 2006; Oberdörster et al. 2002, 2004). The nasal mucosa has a pH close to neutral (Washington 2000) and the olfactory mucosa does not contain phagocytic cells under normal conditions (Ross and Romrell 1989).

Particulate material is deposited in the alveoli via diffusional processes, as bulk air flow is low or absent in this region (Schlesinger 1995). Two important anatomical features of the alveolar region to consider include: (1) its large surface area of 80–140 m^2 in humans (Scheuch et al. 2006) to facilitate gas exchange; and (2) the large extent of vascularization. At the epithelial surface of the alveolus, there are only a few cell types in a healthy lung with which nanomaterials might interact. The two alveolar epithelial cell types are squamous (Type I) and cuboidal (Type II) in morphology; whereas these two cell types have similar number distributions in the alveolus, the Type I cells cover ∼95% of the alveolar surface (Ross and Romrell 1989). The Type II cells differentiate into Type I cells (Adamson and Bowden 1975) and also proliferate to repair injured areas of the alveolus. Gas exchange occurs through the thin, extended filipodia of the Type I cells, which form *zonulae occludens* (occluding tight junctions) with other Type I cells. The basement membrane of the Type I epithelial cell is continuous with that of the endothelial cells lining the pulmonary capillaries, except for a thin interstitium, so the total thickness through which gases (or nanoparticles) have to travel to reach the blood is 0.36–2.5 μm (Phalen et al. 1995). The pulmonary capillaries are immediately proximal to the alveoli and form a dense, intertwining network in the parenchymal region of the lung. The third cell type with which NPs might interact is the alveolar macrophage, which has the principal job of removing cellular debris and other particulate material. The brush cell is another cell type in the alveolus, often located at the bronchoalveolar duct junction, that has also been described by some anatomists; however, it is rare and its function is largely unknown (Reid et al. 2005).

Like skin cells, tight junctions exist between lung epithelial cells; however, the barrier function is different than in skin. Selective permeability and active transport of ions through tight junctions give rise to a transepithelial potential difference giving lung mucosa an anionic charge. Hence, like the skin, nanoparticles containing high positive surface potential might experience stronger interactions with the lung mucosa and be more prone to interaction with cell membranes if the particles reach the cell surface (Grenha et al. 2005; Yacobi et al. 2008). However, a cautionary note should be added: many studies that investigate surface charge as a determinant of cellular interactions with NPs have been done with cultured cells. The lipids and proteins present in lining fluid in vivo also have charged surfaces, thus adding to the complexity of this issue.

The fluid system that lines the airways and alveoli functions in maintaining the normal physiology of the respiratory tract (e.g., mucociliary clearance, patency of alveoli) as well as in host defense. The composition of the fluid varies, however, in the different regions of the lung. The conducting airways are lined with a complex mixture of mucous substances and aqueous components that varies in depth from ~5 to 100 μm (Ng et al. 2004) and also has some phagocytic cells. These components coordinate to affect host defense and maintain mucociliary clearance. The alveolar lining fluid consists of surfactants and an overlying aqueous phase. The pulmonary surfactant is ~90% lipids and 10% proteins. The main physiological role of surfactants is to keep both the alveoli and bronchioles patent during respiration. The lipid component is composed largely of disaturated dipalmitoylphosphatidylcholine and phosphatidylglycerol with smaller amounts of cholesterol. There are four surfactant proteins (SPs) associated with the lipid layer: SP-A, SP-B, SP-C, and SP-D; they are secreted by Type II alveolar epithelial and Clara cells (Griese 1999). The alveolar lining fluid also contains plasma-derived proteins (e.g., albumin, transferrin, immunoglobulins) that are critical to host defense functions (Kim and Malik 2003), as mucociliary clearance is absent in this region of the lung.

3.4.2 Lessons from the Study of Ambient Air and Industrial Ultrafine Particles

People are exposed on a daily basis to nanosized particles, namely the ultrafine particles (UFP) that are present with oxidant gases in ambient air pollution aerosols. Strictly in terms of size, ultrafine and nanosized particles are the same (i.e. <100 nm). However, UFP are heterogeneous in their chemistry, are usually present in air as agglomerates, and are produced via combustion-related processes from human activity. In this chapter, we define agglomeration as the reversible attachment of groups of nanoparticles; aggregated nanoparticles, then, are ones that cannot be broken apart except under the harshest of conditions that are not likely to be present in biological tissues (except, perhaps, phagolysosomes of activated macrophages). More recent physiochemical characterizations of ambient UFP have also revealed that they are not necessarily solid particles, but are comprised to a significant degree of volatile and semi-volatile organic species (Kittelson et al. 2004). Despite these differences, some important lessons that are likely to be applicable to nanoparticles have been learned by studying the health effects of ambient air particulate matter and, in particular, UFP.

One lesson from the study of ambient particulate matter is that an efficient clearance mechanism for UFP is their agglomeration onto larger particles (heterogeneous). Agglomeration processes within the UFP fraction (homogeneous) are less efficient. This could impact the biological responses to nanoparticles in that agglomerated materials could act as a sink for singlet nanoparticles, thus potentially reducing the ability of cells and tissues to interact with smaller particles. However, very little is known from studying either UFP or nanoparticles about the behavior of agglomerated or aggregated particles in tissues.

Another important lesson, as mentioned above, is that the deposition of nano-sized particles in the alveolar region is size-dependent, with a peak ~20 nm (ICRP 1994). Asthmatics deposit even higher amounts of UFP when at rest as compared to healthy individuals (Frampton et al. 2004); possibly related to airway remodeling and/or changes in breathing patterns. In addition, alveolar macrophages do not efficiently take up singlet UFP (Hahn et al. 1977; Ahsan et al. 2002). Hence, as mentioned in Section 3.2.2, alveolar macrophages are not likely to clear nano-sized particles unless they are aggregated or agglomerated. These two factors lead to the potential for increased interaction of nanoparticles with epithelial structures in the alveoli, increased retention in the lung, and their subsequent passage from the epithelial surface through the interstitium and into the microvascular circulation of the lungs in a process called translocation. Several studies have now shown that inhaled insoluble UFP are translocated in small, but significant levels to tissues outside of the respiratory tract, e.g., the liver (Kreyling et al. 2002; Oberdörster et al. 2002). More recent studies have also shown the olfactory bulb to be a target for UFP translocation. Studies with inhaled insoluble Mn oxide and ^{13}C UFP showed that about 11% and 20%, respectively, of the inhaled amount that deposited in the nose traveled to the olfactory bulb (Elder et al. 2006; Oberdörster et al. 2004). Whether or not translocation away from the original site of exposure occurs for all tissues and the extent to which the process is dependent on key physiochemical properties of the nanoparticles (Section 3.2) are issues being addressed by current research.

Subchronic multispecies inhalation studies with aerosolized carbon black and TiO$_2$ showed impaired clearance, more persistent and severe inflammation, more severe lung epithelial cell proliferation, and greater oxidant stress in animals exposed to nanosized, high surface area particles as compared to larger ones with smaller surface areas (Bermudez et al. 2002, 2004; Carter et al. 2006; Elder et al. 2005). In addition to these in vivo studies, many years of epidemiological, in vivo, and in vitro research with ambient air and industrially relevant particulate matter support the hypothesis that small particles with high surface areas produce greater adverse effects as compared to larger particles with the same chemistry (Li et al. 1996; Oberdörster et al. 1994, 2000; Peters et al. 1997; Timonen et al. 2005). Studies with UFP have also shown that oxidative stress is a consequence of exposure, as evidenced by enhanced oxidant release and activation of bronchoalveolar lavage (BAL) inflammatory cells, increased intracellular oxidant production, oxidative DNA damage, mitochondrial damage, heme oxygenase -1 protein induction, cellular thiol depletion, and proinflammatory cytokine induction (Elder et al. 2000, 2004; Li et al. 1996, 2003; Xia et al. 2006). Health status also significantly affects the response to inhaled particulate matter, including UFP. Epidemiological studies show that age and cardiopulmonary disease determine the outcome of exposure to ambient particulate matter in terms of human morbidity and mortality (Bateson and Schwartz 2004; Zanobetti and Schwartz 2001, 2002). Animal studies have supported the findings in human cohorts (Elder et al. 2004, 2007; Kodavanti et al. 2002; Lambert et al. 2003). Similar effects are likely to be observed should the respiratory tract be a target for intentional or accidental nanoparticle exposure.

UFP have also been shown to have effects outside of the respiratory tract. Such effects include enhanced venous thrombus and atherosclerotic lesion formation, alterations in circulating thrombin–anti-thrombin complex and fibrinogen levels, inflammatory mediator production in cortical neurons and olfactory bulb, and alterations in the heart rate and heart rate variability (Campbell et al. 2005; Elder et al. 2004, 2006, 2007; Künzli et al. 2005; Nemmar et al. 2002; Silva et al. 2005; Timonen et al. 2005). These extrapulmonary effects that have been observed following exposure to UFP may be due either to direct transport of particles to other tissues or to the action of soluble mediators (or a combination of these two processes). The preceding discussion (Section 3.4.1) of respiratory tract anatomy describes how direct translocation might be possible.

Studies with diverse nanomaterials suggest that oxidative stress is a likely outcome of exposure in cultured cells and animals. Although still somewhat controversial, Nel et al. (2006) proposed a categorization of oxidative stress responses in terms of severity. The least severe oxidative stress is that which leads to antioxidant gene induction via the transcription factor, nuclear factor E2-related factor 2. Followed by that is nuclear factor kappa B or activator protein-1-mediated cytokine and chemokine gene induction and, finally, cell death by either apoptotic or necrotic pathways involving mitochondrial dysfunction. Early studies, from what is now ancient toxicological history for nanoparticles, showed that nC_{60} fullerenes caused lipid oxidation in brain tissues from fish (Oberdörster 2004). However, subsequent in vitro studies showed that surface derivatization to make the aggregated fullerenes more water soluble also made them less toxic (Sayes et al. 2004). One of the most heavily studied engineered nanosized respiratory tract hazards is the carbon nanotube (CNT). Early intratracheal instillation studies reported mortality and granuloma formation in lung tissue (Lam et al. 2004; Warheit et al. 2004); however, these effects were largely due to the instilled doses, the dose rates, and the agglomerate state of the CNTs. Later studies at lower doses performed by Shvedova et al. (2005) showed that granulomas formed in those regions where visible aggregates were present in lung tissue; nevertheless, there were fibrotic changes in regions where aggregates were not seen at the limits of light microscopy. A recent study (Li et al. 2007) with inhaled multi-walled CNTs delivered at doses near what has been observed in the workplace reported only moderate epithelial proliferation and alveolar wall thickening retained nanotube agglomerates. Taken together, these data suggest that dose rate and agglomerate state play significant roles in determining the outcome following respiratory tract exposures to CNTs. These issues are also likely to be important with other nanomaterial types.

3.5 Conclusions and Future Directions

With the surging nanotechnology industry, the likelihood of intentional consumer and unintended worker-related skin and lung exposures to various types of nanomaterials is assured (Baron et al. 2002). From existing literature, there is clear evidence that polymeric, rigid metal and semiconductor nanomaterials can passively breech epithelial barriers. For skin, mechanical flexing can facilitate penetration of large micron-sized particles and, for both skin and lung, the health status will affect barrier function. Nanoparticle toxicology is, however, an emerging field and inconsistencies in the published literature exist, as is exemplified by the data on metal oxide nanomaterials permeation through skin and effects of CNTs in the respiratory tract. Inconsistencies should be anticipated as there is currently no standardized set of tests by which nanoparticles toxicity can be determined. For example, the nature of the vehicles used to apply nanomaterials to skin, the methods used for respiratory tract exposures, the doses, the nanoparticle themselves, and their purity vary tremendously. There may be nanoparticle physiochemical factors other than size, surface charge, and hydrophobicity, such as mechanical properties, that are also important to consider in correlating nanomaterial penetration data.

Therefore, the question of nanomaterial toxicity resulting from unintended epithelial permeation remains open. In vitro cytotoxicity studies clearly indicate that nanomaterials are toxic to skin and lung cells under certain conditions. The relevance of these results is difficult to extrapolate, as there is a presumption of epithelial permeation. Quantification of the latter from existing studies suggests that permeation as a percent of total particles applied is low for skin and the respiratory tract; however, exact mass quantification is difficult. Moreover, particularly for skin, many studies are conducted using ex vivo tissue, the results of which should be viewed with caution. Ex vivo tissue does offer a convenient test platform to probe permeation; however, the tissue is dying. Circulatory systems are absent and normal cellular functions, including immune responses and differentiation, are maintained for only a few hours (<24 hrs). Ex vivo skin pH, sweat and sebum production, may differ considerably from in vivo tissue. Such factors are critically important when considering the effect that salt, acidity, and surface tension have on nanoparticles to resist agglomeration or degradation. Clearly, in vivo studies offer the best opportunity to investigate nanomaterial penetration and translocation mechanisms. Unfortunately, limited published data are available at this time, but certainly will be the focus of future studies.

References

Adamson IY, Bowden DH (1975) Derivation of type I epithelium from type II cells in the developing rat lung. Lab Invest 32(6):736–745.

Åkerman ME, Chan WCW, Laakkonen P, Bhatia SN, Ruoslahti E (2002) Nanocrystal targeting in vivo. PNAS 99:12617–12621.

Alvarez-Roma R, Naika A, Kaliaa YN, Guy RH, Fessi H (2004) Skin penetration and distribution of polymeric nanoparticles. J Control Release 99(1):53–62.

Anderson JM (2001) Molecular structure of tight junctions and their role in epithelial transport. News Physiol Sci 16:126–130.

Ahsan F, Rivas IP, Khan MA, Suárez-Torres AI (2002) Targeting to macrophages: role of physicochemical properties of particulate carriers – liposomes and microspheres – on the phagocytosis by macrophages. J Control Release 79:29–40.

Auffan M, Decome L, Rose J, Orsiere T, De Meo M, Brisis V, Chaneac C, Olivi L, Berge-Lefranc J-L, Botta A, Wiesner MR, Bottero J-Y (2006) In vitro interactions between DMSA-coated maghemite nanoparticles and human fibroblasts: a physicochemical and cyto-genotoxical study. Environ Sci Technol 40(14):4367–4373.

Badea I, Wettig S, Verrall R, Foldvari, M (2007) Topical non-invasive gene delivery using gemini nanoparticles in interferon-γ-deficient mice. Eur J Pharm Biopharm 65:414–422.

Bateson TF, Schwartz J (2004) Who is sensitive to the effects of particulate air pollution on mortality? A case-crossover analysis of effect modifiers. Epidemiology 15(2):143–149.

Bailey M.R. (1994) The new ICRP model for the respiratory tract. Radiation Protection Dosimetry 53:107–114.

Baroli B, Ennas MG, Loffredo F, Isola M, Pinna R, Lopez-Quintela MA (2007) Penetration of metallic nanoparticles in human full-thickness skin, J Invest Dermatol 127(7):1701–1712.

Barbero AM, Frasch HF (2006) Transcellular route of diffusion through stratum corneum: results from finite element models, J Pharm Sci 95(10):2186–2194.

Baron PA, Maynard AD, Foley M (2002) Evaluation of aerosol release during the handling of unrefined single walled carbon nanotube material. National Institute for Occupational Safety and Health, Cincinnati (OH), NIOSH report:DART-02-191.

Bermudez E, Mangum JB, Asgharian B, Wong BA, Reverdy EE, Janszen DB, Hext PM, Warheit DB, Everitt JI. (2002) Long-term pulmonary responses of three laboratory rodent species to subchronic inhalation of pigmentary titanium dioxide. Toxicol Sci 70:86–97.

Bermudez E, Mangum JB, Wong BA, Asgharian B, Hext PM, Warheit DB, Everitt JI (2004) Pulmonary Responses of Rats, Mice, and Hamsters to Subchronic Inhalation of Ultrafine Titanium Dioxide Particles. Toxicol Sci 77:347–357.

Berry CC, Wells S, Charles S, Curtis AS (2003) Dextran or albumin derivatised iron oxide nanoparticles: influence on fibroblasts in vitro. Biomaterials 24(25):4551–4557.

Berry CC, Charles S, Wells S, Dalby MJ, Curtis AS (2004) The influence of transferrin stabilized magnetic nanoparticles on human dermal fibroblasts in culture. Int J Pharm 269(1): 211–225.

Bodian D, Howe HA (1941a) Experimental studies on intraneural spread of poliomyelitis virus, *Bull Johns Hopkins Hosp* 68:248–267.

Bodian D, Howe HA (1941b) The rate of progression of poliomyelitis virus in nerves. *Bull Johns Hopkins Hosp* 69:79–85.

The Royal Society and the Royal Academy of Engineering (2004) Nanoscience and nanotechnologies: opportunities and uncertainties. The Royal Society.

Brouxhon S, Kyranides S, O'Banion MK, Johnson R, Pearce DA, MGrath K, Erdle B, Scott G, Schneider S, VanBuskirk J, Pentland AP (2007) Sequential downregulation of E-cadherin with squamous cell carcinoma progression: loss of E-cadherin via a prostaglandin E_2–EP_2 dependent mechanism. Cancer Res (in Press).

Campbell A, Oldham M, Becaria A, Bondy SC, Meacher D, Sioutas C, Misra C, Mendez LB, Kleinman M (2005) Particulate matter in polluted air may increase biomarkers of inflammation in mouse brain. Neurotoxicology 26:133–140.

Carter JM, Corson N, Driscoll KE, Elder A, Finkelstein JN, Harkema JR, Gelein R, Wade-Mercer P, Nguyen K, Oberdörster G (2006) A comparative dose-related response of several key pro- and antiinflammatory mediators in the lungs of rats, mice, and hamsters after subchronic inhalation of carbon black. J Occup Environ Med 48:1265–1278.

Cedervall T, Lynch I, Lindman S, Berggård T, Thulin E, Nilsson H, Dawson KA, Linse S (2007) Understanding the nanoparticle–protein corona using methods to quantify exchange rates and affinities of proteins for nanoparticles. PNAS 104:2050–2055.

Chang E, Yu WW, Colvin VL, Drezek RJ (2005) Quantifying the influence of surface coatings on quantum dot uptake in cells. J Biomed Nanotechnol 1:397–401.

Chang E, Thekkek N, Yu WW, Colvin VL, Drezek R (2006) Evaluation of QDot cytotoxicity based on intracellular uptake. Small 2(12):1412–1417.

Choi HS, Liu W, Misra P, Tanaka E, Zimmer JP, Ipe BI, Bawendi MG, Frangioni JV (2007) Renal clearance of quantum dots. Nature biotechnology 25:1165–1170.

Cross SE, Innes B, Roberts MS, Tsuzuki T, Robertson TA, McCormick P (2007) Human skin penetration of sunscreen nanoparticles. Skin Pharmacol Physiol 20(3):148–154.

Cui Z, Mumper RJ (2001) Chitosan-based nanoparticles for topical genetic immunization. J Control Release 75(3):409–419.

Curtis J, Greenberg M, Kester J, Phillips S, Krieger G (2006) Nanotechnology and nanotoxicology: a primer for clinicians. Toxicol Rev 25(4):245–260.

Dayan N (2005) Pathways for skin penetration. Cosmetics & Toiletries Magazine 120(6):67–76.

DeLorenzo, AJD (1970) The olfactory neuron and the blood-brain barrier. In: Wolstenholme GEW, Knight J (eds) Taste and smell in vertebrates. J&A Churchill, London, pp 151–176.

Dick IP, Scott RC (1992) Pig ear skin as an in-vitro model for human skin permeability. J Pharm Pharmacol 44(8):640–645.

Ding L, Stilwell J, Zhang T, Omeed Elboudwarej O, Huijian Jiang H, Selegue JP, Cooke PA, Gray JW, Chen FF (2005) Molecular characterization of the cytotoxic mechanism of multiwall carbon nanotube and nano-onions on human skin fibroblast. Nano Lett 5(12): 2448–2464.

Derfus AM, Chan WCW, Bhatia SN (2004) Probing the cytotoxicity of Qdots. Nano Lett 4(1): 11–18.

Destree C, Ghijsen J, Nagy JB (2007) Preparation of organic nanoparticles using microemulsions: their potential use in transdermal delivery. Langmuir 23(4):1965–1973.

Eedy DJ (1996) Carbon-fibre-induced airborne irritant contact dermatitis. Contact Dermatitis 35(6):362–363.

Elder A, Gelein R, Finkelstein J, Phipps R, Frampton M, Utell M, Topham D, Kittelson D, Watts W, Hopke P, Jeong C-H, Kim E, Liu W, Zhao W, Zhou L, Vincent R, Kumarathasan P, Oberdörster G (2004) On-road exposure to highway aerosols. 2. Exposures of aged, compromised rats. Inhal Toxicol 16:41–53.

Elder A, Gelein R, Finkelstein JN, Driscoll KE, Harkema J, Oberdörster G (2005) Effects of sub-chronically inhaled carbon black in three species. I. Retention kinetics, lung inflammation, and histopathology. Toxicol Sci 88:614–629.

Elder A, Gelein R, Silva V, Feikert T, Opanashuk L, Carter J, Potter R, Maynard A, Ito Y, Finkelstein J, Oberdörster G (2006) Translocation of inhaled ultrafine manganese oxide particles to the central nervous system. Environ. Health Perspect 114:1172–1178.

Elder A, Couderc J-P, Gelein R, Eberly S, Cox C, Xia X, Zareba W, Hopke P, Watts W, Kittelson D, Frampton M, Utell M, Oberdörster G (2007) Effects of on-road highway aerosol exposures on autonomic responses in aged, spontaneously hypertensive rats. Inhal Toxicol 19: 1–12.

Elder ACP, Gelein R, Finkelstein JN, Cox C, Oberdörster G (2000) The pulmonary inflammatory response to inhaled ultrafine particles is modified by age, ozone exposure, and bacterial toxin. Inhal Toxicol 12:227–246.

Fien SM, Oseroff AR (2007) Photodynamic therapy for non-melanoma skin cancer. J Natl Compr Cancer Network 5(5):531–540.

Fischer HC, Liu L, Pang KS, Chan WCW (2006) Pharmacokinetics of nanoscale quantum dots: in vivo distribution, sequestration, and clearance in the rat. Adv Funct Mater 16(10): 1299–1305.

Frampton MW, Utell MJ, Zareba W, Oberdörster G, Cox C, Huang LS, Morrow PE, Lee FE, Chalupa D, Frasier LM, Speers DM, Stewart JC (2004) Effects of exposure to ultrafine carbon particles in healthy subjects and subjects with asthma. Res Rep Health Eff Inst 126:1–47.

Gamer AO, Leibold E, van Ravenzwaay B (2006) The in vitro absorption of microfine zinc oxide and titanium dioxide through porcine skin. Toxicol in Vitro 20(3):301–307.

Gbadamosi HJ, Hunter AC, Moghimi SM (2002) PEGylation of microspheres generates a heterogeneous population of particles with differential surface characteristics and biological performance. FEBS Lett, 532(3):338–344.

Ghadially R, Brown BE, Sequeira-Martin SM, Feingold KR, Elias PM (1995) The aged epidermal permeability barrier: structural, functional, and lipid biochemical abnormalities in humans and a senescent murine model. J Clin Invest 95:2281–2290.

Ghitescu L, Fixman A (1984) Surface charge distribution on the endothelial cells of liver sinusoids. J Cell Biol, 99:639–647.

Gopee NV, Roberts DW, Webb P, Cozart CR, Siitonen PH, Warbritton AR, Yu WW, Colvin VL, Walker NJ, Howard PC (2007) Migration of intradermally injected quantum dots to sentinel organs in mice. Toxicol Sci 98(1):249–257.

Grenha A, Seijo B, Remuñán-López C (2005) Microencapsulated chitosan nanoparticles for lung protein delivery. Eur J Pharm Sci 25(4–5):427–437.

Griese M (1999) Pulmonary surfactant in health and human lung diseases: state of the art. Eur Resp J 13:1455–1476.

Gupta AK, Gupta, M (2005) Cytotoxicity suppression and cellular uptake enhancement of surface modified magnetic nanoparticles. Biomaterials 26:1563–1573.

Hahn FF, Newton GJ, Bryant PL (1977) In vitro phagocytosis of respirable-sized monodisperse particles by alveolar macrophages. In Sanders CL, Schneider RP, Dagle GE, Ragen HA (eds) Pulmonary macrophages and epithelial cells, vol 43. Technical Information Center, Energy Research and Development Administration, Oak Ridge, pp 424–435.

Hardman R (2006) A toxicologic review of quantum dots: toxicity depends on physicochemical and environmental factors. Environ Health 114(2):165–172.

Holmberg SB, Forssell-Aronsson E, Gretarsdottir J, Jacobsson L, Rippe B, Hafstrom L (1990) Vascular clearance by the reticuloendothelial system–measurements using two different-sized albumin colloids. Scand J Clin Lab Invest 50(8):865–871.

Honeywell-Nguyen PL, Gooris GS, Bouwstra JA (2004) Quantitative assessment of the transport of elastic and rigid vesicle components and a model drug from the vesicle formulation into human skin in vivo. JID 123:902–910.

Hoshino A, Fujioka K, Oku T, Suga M, Sasaki YF, Ohta T, Yasuhara M, Suzuki K, Yamamoto K (2004) Physicochemical properties and cellular toxicity of nanocrystal quantum dots depend on their surface modification. Nano Lett 4:2163–2169.

Hostynek JJ, Maibach HI (2004) Copper hypersensitivity: dermatologic aspects. Dermatol Ther 17(4):328–333.

Hostynek JJ, Dreher F, Nakada T, Schwindt D, Anigbogu A, Maibach HI (2001) Human stratum corneum adsorption of nickel salts. Investigation of depth profiles by tape stripping in vivo. Acta Derm Venereol Suppl 212:11–18.

Hueber F, Wepierre J, Schaefer H (1992) Role of transepidermal and transfollicular routes in percutaneous absorption of hydrocortisone and testosterone: in vivo study in the hairless rat. Skin Pharmacol 5:99–107.

International Committee on Radiological Protection (1994) Human Respiratory Tract Model for Radiological Protection. A Report of Committee 2 of the ICRP, Pergamon Press, Oxford.

Illel B (1997) Formulation for transfollicular drug administration: some recent advances. Crit Rev Ther Drug Carrier Syst 14(3):207–219.

Ipe BI, Lehnig M, Niemeyer CM (2005) On the generation of free radical species from quantum dots. Small 1:706–709.

Javier AM, Kreft O, Alberola AP, Kirchner C, Zebli B, Susha AS, Horn E, Kempter S, Skirtach AG, Rogach AL, Rädler J, Sukhorukov GB, Benoit M, Parak WJ (2006) Combined atomic force microscopy and optical microscopy measurements as a method to investigate particle uptake by cells. Small 2(3):394–400.

Jiang JS, Chu AW, Lu ZF, Pan MH, Che DF, Shou XJ (2007) Ultraviolet B-induced alterations of the skin barrier and epidermal calcium gradient. Exp Dermatol 16:985–992.

Junquiera CL, Carneiro J, Kelley RO (1992) Basic histology, 7th edn. Appleton and Lange, Englewood Cliffs NJ.

Kim MK, Choi SY, Byun HJ, Huh CH, Park KC, Patel RA, Shinn AH, Youn SW (2006) Comparison of sebum secretion, skin type, pH in humans with and without acne, Arch Dermatol Res 298:113–119.

Kim KJ, Malik AB (2003) Protein transport across the lung epithelial barrier. Am J Physiol 284(2):L247–L259.

Kittelson DB, Watts WF, Johnson JP, Remerowski ML, Ische EE, Oberdörster G, Gelein RM, Elder A, Hopke PK (2004) On-road exposure to highway aerosols. 1. Aerosol and gas measurements. Inhal Toxicol 16:31–39.

Klein J (2007) Probing the interactions of proteins and nanoparticles, PNAS 104:2029–2030.

Kodavanti UP, Schladweiler MC, Ledbetter AD, Hauser R, Christiani DC, McGee J, Richards JR, Costa DL (2002) Temporal association between pulmonary and systemic effects of particulate matter in healthy and cardiovascular compromised rats. J Toxicol Environ Health 65(20): 1545–1569.

Kohli AK, Aplar HO (2004) Potential use of nanoparticles for transcutaneous vaccine delivery: effect of particle size and charge. Int J Pharm 275:13–17.

Kreyling WG, Semmler M, Erbe F, Mayer P, Takenaka S, Schulz H (2002) Translocation of ultrafine insoluble iridium particles from lung epithelium to extrapulmonary organs is size dependent but very low. J Toxicol Environ Health 65:1513–1530.

Künzli N, Jerret M, Mack WJ, Beckerman B, LaBree L, Gilliland F, Thomas D, Peters J, Hodis HN (2005) Ambient air pollution and atherosclerosis in Los Angeles. Environ Health Perspect 113:201–206.

Lademann J, Weigmann H, Rickmeyer C, Barthelmes H, Schaefer H, Mueller G, Sterry W (1999) Penetration of titanium dioxide microparticles in the Horny Layer and the Follicular Orifice. Skin Pharmacol Appl Skin Physiol 12:247–256.

Lademann J, Richter H, Schaefer UF, Blume-Peytavi U, Teichmann A, Otberg N, Sterry W (2006a) Hair follicles – A long-term reservoir for drug delivery. *Skin Pharmacol Physiol* 19(4): 232–236.

Lademann J, Ilgevicius A, Zurbau O, Liess HD, Schanzer S, Weigmann HJ, Antoniou C, Pelchrzim RV, Sterry W (2006b) Penetration studies of topically applied substances: optical determination of the amount of stratum corneum removed by tape stripping. J Biomed Optics 11(5): 054026–1 to 054026–6.

Lademann J, Richter H, Teichmann A, Otberg N, Blume-Peytavi U, Luengo J, Weiss B, Schaefer UF, Lehr CM, Wepf R, Sterry W (2007) Nanoparticles – An efficient carrier for drug delivery into the hair follicles. Eur J Pharm Biopharm, 66:159–164.

Lam CW, James JT, McCluskey R, Hunter RL (2004) Pulmonary toxicity of single-wall carbon nanotubes in mice 7 and 90 days after intratracheal instillation. Toxicol Sci 77:126–134.

Lambert AL, Mangum JB, DeLorme MP, Everitt JI (2003) Ultrafine carbon black particles enhance respiratory syncytial virus-induced airway reactivity, pulmonary inflammation, and chemokine expression. Toxicol Sci 72:339–346.

Lee S, Lee J, Choi YW (2007) Skin permeation enhancement of ascorbyl palmitate by liposomal hydrogel (lipogel) formulation and electrical assistance. Biol & Pharm Bull 40(2):393–396.

Li L, Hamilton RF Jr, Kirichenko A, Holian A (1996) 4-hydroxynonenal-induced cell death in murine alveolar macrophages. Toxicol Appl Pharmacol 139:135–143.

Li N, Sioutas C, Cho A, Schmitz D, Misra C, Sempf J, Wang M, Oberley T, Froines J, Nel A (2003) Ultrafine particulate pollutants induce oxidative stress and mitochondrial damage. *Environ Health Perspect* 111:455–460.

Li JG, Li WX, Xu JY, Cai XO, Liu RL, Li YJ, Zhao QF, Li QN (2007) Comparative study of pathological lesions induced by multiwalled carbon nanotubes in lungs of mice by intratracheal instillation and inhalation. Environ Toxicol 22(4):415–421.

Liu J, Hu W, Chen H, Ni Q, Xu H, Yang X (2006) Isotretinoin-loaded solid lipid nanoparticles with skin targeting for topical delivery. Int J Pharm 328(2):191–195.

Lovrić J, Cho SJ, Winnik FM, Maysinger D (2005a) Unmodified cadmium telluride quantum dots induce reactive oxygen species formation leading to multiple organelle damage and cell death. Chem Biol 12:1227–1234.

Lovrić J, Bazzi HS, Cuie Y, Fortin GRA, Winnik FM, Maysinger D (2005b) Differences in subcellular distribution and toxicity of green and red emitting CdTe quantum dots. J Mol Med 83:377–385.

Lutz O, Meraihi Z, Mura JL, Frey A, Riess GH, Bach AC (1989) Fat emulsion particle size: influence on the clearance rate and the tissue lipolytic activity. Am J Clin Nutr 50(6):1370–1381.

Marro D, Guy RH, Delgado-Charro MB (2001) Characterizaton of the iontophoretic permselectivity properties of human and pig skin. J Control Release 70:213–217.

Manna SK, Sarkar S, Barr J, Wise K, Barrera EV, Jejelowo O, Rice-Ficht AC, Ramesh GT (2005) Single walled carbon nanotube induces oxidative stress and activates nuclear transcription factor kappa B in human keratinocyte cells. Nano Lett 5(9):1676–1684.

Meidan VM, Bonner MC, Michniak BB (2006) Transfollicular drug delivery – Is it a reality? Int J Pharm, 306(1–2):1–14.

Menon GK, Feingold KR, Elias PM (1992) Lamellar body secretory response to barrier disruption. J Invest Dermatol, 98:279–289.

Mortensen L, Pentland AP, Oberdorstor G, DeLouise LA (2007) In vivo study of quantum dots penetration through skin, 2007 preliminary data.

Monterio-Riviere NA, Nemanich RJ, Inman AO, Wang YY, Riviere JE (2005a) Multi-walled carbon nanotube interactions with human epidermal keratinocytes. Toxicol Lett 155:377–384.

Monteiro-Riviere NA, Inman AO, Wang YY, Nemanich RJ (2005b) Surfactant effects on carbon nanotube interactions with human keratinocytes. Nanomed: Nanotechnol, Biol, Med 1: 293–299.

Morones JR, Elechiguerra JL, Camacho A, Holt K, Kouri JB, Ramirez JT, Yacaman MJ (2005) The bactericidal effect of silver nanoparticles. Nanotechnology 16:2346–2353.

Nel A, Xia T, Mädler L, Li N (2006) Toxic potential of materials at the nanolevel. Science 311(3):622–627.

Ng AW, Bidani A, Heming TA (2004) Innate host defense of the lung: effects of lung-lining fluid pH. Lung 182:297–317.

Nemmar A, Hoylaerts M, Hoet PHM, Dinsdale D, Smith T, Xu H, Vermylen J, Nemery, B (2002) Ultrafine particles affect experimental thrombosis in an in vivo hamster model. Am J Respir Crit Care Med 166(7):598–1004.

Oberdörster E (2004) Manufactured nanomaterials (Fullerenes, C60) induce oxidative stress in the brain of juvenile largemouth bass. Environ Health Perspect 112:1058–1062.

Oberdörster G, Ferin J, Lehnert B (1994) Correlation between particle size, in vivo particle persistence, and lung injury. Environ Health Perspect 102:173–179.

Oberdörster G, Finkelstein JN, Johnston C, Gelein R, Cox C, Baggs R, Elder ACP (2000) Acute pulmonary effects of ultrafine particles in rats and mice. Health Effects Institute, Cambridge, Report 96, pp. 1–74.

Oberdörster G, Sharp Z, Atudorei V, Elder A, Gelein R, Lunts A, Kreyling W, Cox C (2002) Extrapulmonary translocation of ultrafine carbon particles following whole-body inhalation exposure of rats. J Toxicol Environ Health 65:1531–1543.

Oberdörster G, Sharp Z, Atudorei V, Elder A, Gelein R, Kreyling W, Cox C (2004) Translocation of inhaled ultrafine particles to the brain. Inhal Toxicol 16:437–445.

Oberdörster G, Oberdörster E, Oberdörster J (2005) Nanotoxicology: an emerging discipline evolving from studies of ultrafine particles. Environ Health Persp 113(71):823–839.

Pentland AP, Schoggins JW, Scott GA, Khan KNM, Rujing Han R (1999) Reduction of UV-induced skin tumors in hairless mice by selective COX-2 inhibition. Carcinogenesis 20(10):1939–1944.

Parisel C, Saffar L, Gattegno L, Andre V, Abdul-Malak N, Perrier E, Letourneur D (2003) Interactions of heparin with human skin cells: binding, location, and transdermal penetration. Journal of Biomedical Materials Research Part A 67(2):517–523.

Pernodet N, Fang XH, Sun Y, Bakhtina A, Ramakrishnan A, Sokolov J, Ulman A, Rafailovich M (2006) Adverse effects of citrate/gold nanoparticles on human dermal fibroblasts. Small 2(6):766–773.

Peters A, Wichmann H-E, Tuch T, Heinrich J, Heyder J (1997) Respiratory effects are associated with the number of ultrafine particles. Am J Respir Crit Care Med 155:1376–1383.

Phalen RF, Yeh H, Praasad SB (1995) Morphology of the respiratory tract. In: McClellan RO, Henderson RF (eds) Concepts in inhalation toxicology, 2nd edn. Taylor and Francis, Washington, D.C. pp 129–149.

Proksch E, Brasch J (1997) Influence of epidermal permeability barrier disruption and Langerhans' cell density on allergic contact dermatitis. Acta Derm Venereol 108, 73.

Pulskamp K, Diabate S, Krug HF (2007) Carbon nanotubes show no sign of acute toxicity but induce intracellular reactive oxygen species in dependence on contaminants. Toxicol Lett 168(1):58–74.

Reid L, Meyrick B, Antony VB, Chang LY, Crapo JD, Reynolds HY (2005) The mysterious pulmonary brush cell: a cell in search of a function. Am J Respir Crit Care Med. 172(1):136–139.

Ross MH, Romrell LJ (1989) Histology: a text Williams and Wilkins, Baltimore, MD and atlas, 2nd edn.

Rouse JG, Yang J, Barron AR, Monteiro-Riviere NA (2006) Fullerene-based amino acid nanoparticle interactions with human epidermal keratinocytes. Toxicol in vitro 20(8):1313–1320.

Rouse JG, Yang J, Ryman-Rasmussen JP, Barron AR, Monterio-Riviere NA (2007) Effects of mechanical flexion on the penetration of fullerene amino acid-derivatized peptide nanoparticles through skin. Nano Lett 7(1):155–160.

Ryman-Rasmussen JP, Riviere JE, Montero-Riviere NA (2006) Penetration of intact skin by QDots with diverse physicochemical properties. Toxicol Sci 91(1):159–165.

Ryman-Rasmussen JP, Riviere JE, Monteiro-Riviere NA (2007a) Variables influencing interactions of untargeted QDots with skin cell and identification of biochemical modulators. Nano Lett 7(5):1344–1348.

Ryman-Rasmussen JP, Riviere JE, Montero-Riviere NA (2007b) Surface coatings determine cytotoxicity and irritation potential of quantum dot nanoparticles in epidermal keratinocytes. J Invest Dermatol 127(1):143–153.

Salleh A. (2004) Nano sunblock safety under scrutiny, News in Science ABC Science Online. http://www.abc.net.au/science/news/stories/s1165709.htm

Sayes CM, Fortner J, Lyon D et al (2004) The differential cytotoxicity of water soluble fullerenes. Nano Lett 4:1881–1887.

Sayes CM, Gobin AM, Ausman KD et al (2005) Nano-C60 cytotoxicity is due to lipid peroxidation. Biomaterials 26(36):7587–7595.

Sayes CM, Wahi R, Kurian PA, Liu Y, West JL, Ausman KD, Warheit DB, Colvin VL (2006a) Correlating nanoscale titania structure with toxicity: a cytotoxicity and inflammatory response study with human dermal fibroblasts and human lung epithelial cells. Toxicol Sci 92(1):174–185.

Sayes CM, Liang F, Hudson JL, Mendeza J, Guo W, Beach JM, Moore VC, Doyle CD, West JL, Billups WL, Ausman KD, Colvin VL (2006b) Functionalization density dependence of single-walled carbon nanotubes cytotoxicity in vitro. Toxicol Lett 161:135–142.

Scheuch G, Kohlhaeufl MJ, Brand P, Siekmeier R (2006) Clinical perspectives on pulmonary systemic and macromolecular delivery. Adv Drug Deliv Rev 58:996–1008.

Schulz J, Hohenberg, H, Pflücker F, G ärtner E, Will T, Pfeiffer S, Wcpf R, Wendel V, Gers-Barlag H, Wittern, KP (2002) Distribution of sunscreens on skin. Adv Drug Deliv Rev 54(Suppl. 1):S157–S163.

Schwarz VA, Klein SD, Hornung RÂ, Knochenmuss R, Wyss P, Fink D, Haller U, Walt H (2001) Skin protection for photosensitized patients. Lasers Surg Med 29:252–259.

Schlesinger RB (1995) Deposition and clearance of inhaled particles. In: McClellan RO, Henderson RF (eds) Concepts in inhalation toxicology, 2nd edn. Taylor and Francis, Washington, D.C. pp 191–224.

Shim J, Seok KH, Park WS, Han SH, Kim J, Chang IS (2004) Transdermal delivery of mixnoxidil with block copolymer nanoparticles. J Control Release 97(3):477–484.

Shvedova AA, Castranova V, Kisin ER (2003) Exposure to carbon nanotube material: assessment of nanotube cytotoxicity using human keratinocyte cells. J Toxicol Environ Health A 66(20):1909–1926.

Shvedova AA, Kisin ER, Mercer R, Murray AR, Johnson VJ, Potapovich AI, Tyurina YY, Gorelik O, Arepalli S, Schwegler-Berry D, Hubbs AF, Antonini J, Evans DE, Ku B, Ramsey D, Maynard A, Kagan VE, Castranova V, Baron P (2005) Unusual inflammatory and fibrogenic pulmonary responses to single-walled carbon nanotubes in mice. Am J Physiol 289:L698–L708.

Silva VM, Corson N, Elder A, Oberdörster G (2005) The rat ear vein model for investigating in vivo thrombogenicity of ultrafine particles (UFP). Toxicol Sci 85:983–989.

Sincai M, Argherie D, Ganga D, Bica D, Vekas L (2007) Application of some magnetic nanocompounds in the protection against sun radiation. J Magn Magn Mater 311(1):363–366.

Smith AM, Gao X, Nie S (2004) Quantum dot nanocrystals for in vivo molecular and cellular imaging. Photochem Photobiol 80(3):377–385.

Smart SK, Cassady AI, Lu GQ, Martin DJ (2006) The biocompatibility of carbon nanotubes. Carbon 44:1034–1047.

Tan M, Commens CA, Burnett L, Snitch PJ (1996) A pilot study on the percutaneous absorption of microfine titanium dioxide from sunscreens. Australas J Dermatol 37(4):185–187.

Tian F, Cui D, Schwarz H, Estrada GG, Kobayashi H (2006) Cytotoxicity of single-wall carbon nanotubes on human fibroblasts. Toxicol In Vitro 20(7):1202–1212.

Timonen KL, Vanninen E, de Hartog J, Ibald-Mulli A, Brunekreef B, Gold DR, Heinrich J, Hoek G, Lanki T, Peters A, Tarkiainen T, Tiittanen P, Kreyling W, Pekkanen J (2005) Effects of ultrafine and fine particulate and gaseous air pollution on cardiac autonomic control in subjects with coronary artery disease: the ULTRA study. J Expo Anal Environ Epidemiol 16(4): 332–341.

Tinkle SS, Antonini JM, Rich BA, Roberts JR, Salmen R, DePree K, Adkins EJ (2003) Skin as a route of exposure and sensitization in chronic beryllium disease. Environ Health Perspect 111(9):1202–1208.

Thomas-Ahner JM, Wulff BC, Tober KL, Kusewitt DF, Riggenbach JA, Oberyszyn TM (2007) Gender differences in UVB-induced skin carcinogenesis, inflammation, and DNA damage. Cancer Res 67(7):3468–3474.

Toll R, Jacobi U, Richter H, Lademann J, Schaefer H, Blume-Peytavi U (2004) Penetration profile of microspheres in follicular targeting terminal hair follicles. JID 123:168–176.

Trinchieri G, Kubin M, Bellone G, Cassatella MA (1993) Cytokine cross-talk between phagocytic cells and lymphocytes: relevance for differentiation/activation of phagocytic cells and regulation of adaptive immunity. J Cell Biochem 53(4):301–308.

Tripp CS, Blomme EAG, Chinn KS, Hardy MM, LaCelle P, Pentland AP (2003) Epidermal COX-2 induction following ultraviolet irradiation: suggested mechanism for the roe of COX-2 inhibition in photoprotection. J Invest Dermatol 121:853–861.

Tsay JM, Michalet X (2005) New light on quantum dot cytotoxicity. Chem Biol 12:1159–1161.

Turbill P, Beugeling T, Poot AA (1996) Proteins involved in the Vroman effect during exposure of human blood plasma to glass and polyethylene. Biomaterials 17(13):1279–1287.

Vogt A, Combadiere B, Hadam S, Stieler KM, Lademann J, Schaefer H, Autran B, Sterry W, Blume-Peytavi U (2006) 40 nm, but not 750 or 1,500 nm, nanoparticles enter epidermal CD1a+ cells after transcutaneous application on human skin. J Invest Dermatol 126:1316–1322.

Wagner H, Kosta HH, Lehr CM, Schaefer UF (2003) pH profiles in human skin: influence of two in vitro test systems for drug delivery testing. Euro J Pharma Biopharm 55:57–65.

Warheit DB, Laurence BR, Reed KL, Roach DH, Reynolds GA, Webb TR (2004) Comparative pulmonary toxicity assessment of single-wall carbon nanotubes in rats. Toxicol Sci 77: 117–125.

Washington N, Steele RJC, Jackson SJ, Bush D, Mason J, Gill DA, Pitt K, Rawlins DA (2000) Determination of baseline human nasal pH and the effect of intranasally administered buffers. Int J Pharm 198:139–146.

Weyenberg W, Filev, P, Van den Plas D, Vandervoort J, De Smet K, Sollie P, Ludwig A (2007) Cytotoxicity of submicron emulsions and solid lipid nanoparticles for dermal application. Int J Pharm 337(1–2):291–298.

Witzmann FA, Monteiro-Riviere NA (2006) Multi-walled carbon nanotube exposure alters protein expression in human keratinocytes. Nanomed: Nanotechnol, Biol, Med 2:158–168.

Wissing SA, Muller RH (2002) The influence of the crystallinity of lipid nanoparticles on their occlusive properties, Int J Pharm 242:377–379.

Xia T, Kovochich M, Brant J, Hotze M, Sempf J, Oberley T, Sioutas C, Yeh JI, Wiesner MR, Nel AE (2006) Comparison of the abilities of ambient and manufactured nanoparticles to induce cellular toxicity according to an oxidative stress paradigm. Nano Lett 6:1794–1807.

Yacobi NR, Demaio L, Xie J, Hamm-Alvarez SF, Borok Z, Kim KJ, Crandall ED (2008) Polystyrene nanoparticle trafficking across alveolar epithelium. Nanomed 4(20):139–145.

Zahr AS, Davis CA, Pishko MV (2006) Macrophage uptake of core-shell nanoparticles surface modified with poly(ethylene glycol). Langmuir 22:8178–8185.

Zanobetti A, Schwartz J (2001) Are diabetics more susceptible to the health effects of airborne particles? Am J Respir Crit Care Med 164(5):831–833.

Zanobetti A, Schwartz J (2002) Cardiovascular damage by airborne particles: are diabetics more susceptible? Epidemiology 13(5):588–592.

Zhang T, Stilwell JL, Gerion D, Ding L, Elboudwarej O, Cooke PA, Gray JW, Alivisatos AP, Chen FF (2006) Cellular effect of high doses of silica-coated quantum dot profiled with high throughput gene expression. Nano Lett 6(4):800–808.

Chapter 4
Safety and Efficacy of Nano/Micro Materials

Xiaohong Wei, Yong-kyu Lee, Kang Moo Huh, Sungwon Kim, and Kinam Park

Abstract Nano/micro materials have been used in various applications, and drug delivery is one of the areas where nano/micro particles have made differences. Nano/micro particulate delivery systems can be divided into different categories based on several parameters, such as the nature of nanomaterials (inorganic and organic), biodegradability, hydrophilicity, structures, and processing method. Most of the nano/micro materials in drug delivery have been used without careful considerations of potential toxicity and safety issues. The size, surface area, chemistry, solubility, and shape of nano/micro materials all play significant roles in toxicity. It is time to consider potential problems that may result from the unguided use of nano/micro materials. This chapter deals with potential sources of toxicity in the development of various drug delivery systems.

Contents

K. Park (✉)
Department of Biomedical Engineering, BMED Building, 206 S. Intramural Drive, Purdue University, West Lafayette, IN 47907, USA
e-mail: kpark@purdue.edu

T.J. Webster (ed.), *Safety of Nanoparticles*, Nanostructure Science and Technology, DOI 10.1007/978-0-387-78608-7_4, © Springer Science+Business Media, LLC 2009

4.1 Introduction

Nanotechnology is considered to be one of the most important technologies in modern times. Its unique abilities are expected to revise conventional research and development (R&D) models. For example, cosmetics which have ultrafine clays and oil nanoparticles provide customers significantly improved feelings on their skin; plastics which are modified by carbon nanofibers are as strong as steel, yet as light as hair; and clothes composed of nanofibers are not dampened by the rain. All of these improvements in "incremental technology" [1] have so much potential that many countries are investing considerable resources in this area. Although many promising products, such as carbon nanotubes (CNTs), quantum dots (QDs), sculptured thin films, single-electron transistors, and nanofluidic sensors, have been developed, few of these are available in mass quantities for commercial applications. However, current research and future prospects can provide us with a picture that nanotechnology-based products will be commonly available for consumers within the next decade. It is time to consider potential dangers associated with the preparation, manufacturing and application of nanoparticles. Microparticles are also considered here, as there is no clear boundary separating nanoparticles from microparticles.

The safety protocols of using nanoparticles are urgently required, but have not been given much attention to date [2, 3, 4]. In 2006, the International Risk Governance Council (IRGC) surveyed the current situation of the nanotechnology governance [5]. According to this report, survey participants, consisting of governments of eleven countries, eleven industrial organizations, five research organizations, and nine non-government organizations (NGOs), recognized the importance of R&D activities as well as potential benefits resulting from nanotechnology. Nevertheless, most of respondents did not identify the need for any specified national or international regulations for nanotechnology.

With the ever-increasing R&D activities in nanotechnology, results and data indicating the risks of nanoparticles have been accumulating. Thus, appropriate regulatory action is urgently required to protect human health and the environment from potential disasters [6, 7, 8, 9]. This chapter deals with the analysis of advantages and disadvantages of the preparation, manufacturing, and application of nanoparticles. In particular, the details of safety protocols for overcoming these disadvantages are discussed, along with guidance that is now in place.

4.2 Drugs

The selection of drugs for preparing and manufacturing of nano/micro particles is based on pharmacological activities and market needs. This section is focused on the strategy of maintaining the stability of drugs, especially biomolecular drugs, and on characteristics of non-organic nanoparticles.

4.2.1 Biomolecular Drugs

With advances in biotechnology, more and more biomolecular drugs have been developed for mass production. Clinical applications of those biomolecular drugs, however, have been limited due to their poor stability in formulations and short half-lives in blood. Biomolecular drugs, mainly protein drugs, are prone to denaturation by high temperatures, exposure to organic solvents, contact with solid surfaces, and chemical reactions with other molecules, leading to poor stability during manufacturing and storage of nanoparticulate formulations.

The purity of the protein plays an important role in protein stability/instability [10]. The stability of native proteins is more likely affected by the manufacturing process as compared with chemical modification of the proteins. Poly(ethylene glycol) (PEG) has been used widely for chemical modification (known as PEGylation) of proteins [11]. Methoxy-PEG (MPEG) conjugated proteins [12, 13] have been shown to be more stable than their native counterparts. There are several PEGylated proteins currently on the market, including PEG-adenosine deaminase (Adagen®, Enzon), pefilgrastim (Neulasta®, Amgen), PEG-L-asparaginase (Oncaspar®, Enzon), pegvisomant (Somavert®, Pfizer), PEG-α-interferon-2b (PegIntron®, Schering-Plough), and PEG-α-interferon-2a (Pegasys®, Roche). As an alternative to the non-biodegradable monomethoxy-PEG, poly(sialic acid), a naturally occurring and biodegradable polymer, has been used [14, 15]. Poly(sialic acid) modified proteins were shown to have the same ability to increase the circulation half-life of catalase and asparaginase.

To decrease the degradation of proteins resulting from exposure to the interface during water/oil emulsion processes, anhydrous protein powders have been directly added to polymer-containing organic solvents or the solubility of proteins in organic solvents has been increased. A protein drug can be precipitated at its isoelectric point to make it neutral in charge for dissolution in organic solvents [16]. A protein drug can go through a freeze-drying or spray freeze-drying (SFD) process at a pH away from its isoelectric point [17, 18, 19, 20], resulting in an anhydrous form of that protein. The protein solubility can also be increased based on an ion-pairing mechanism. An oppositely charged surfactant is used to bind the protein and give it a neutral hydrophobic surface. Negatively charged surfactants are usually used to neutralize the positively charged protein because of the toxic side effects of the cationic surfactants [21, 22, 23].

Crystallization is an alternative approach to improving protein stability during microencapsulation procedures, storage and delivery [24], because it only involves a one-step process and results in high purity proteins. Crystalline protein particles are even reported to be more active, stable, and acceptable than their spray-dried amorphous forms [25, 26]. However, few crystalline forms of proteins, especially glycoproteins, have been used as active pharmaceutical ingredients because most proteins are too large and flexible to be crystallized [27].

4.2.2 Inorganic Drugs

4.2.2.1 Magnetic Nanomaterials

Nanoparticles that possess magnetic properties have been extensively investigated as a useful tool for improving the quality of magnetic resonance imaging (MRI), hyperthermic treatment for malignant cells and targeted drug delivery [28]. Iron-containing nanomaterials are controlled by remote magnetic fields, and can be coated with various marker molecules or anti-cancer drugs for targeting within the body. Although neither iron oxide nanoparticle alone nor the coating material alone is known to be toxic, combining the two to create water-soluble nanomaterials produces a completely different effect. They can be toxic to nerve cells and encumber formation of their signal-transmitting extensions [29].

4.2.2.2 Titanium Dioxide Nanomaterials

Titanium dioxide (TiO_2), a noncombustible and odorless white powder, naturally exists in minerals like anatase, rutile, and brookite. It is widely used as a white pigment for paints, paper, plastics, ceramics, for example. TiO_2 is also used as a food additive, such as in toothpaste and capsules, and the Food and Drug Administration (FDA) established a regulation for TiO_2 as the color additive for food. Federal Regulations of the US government regulates the quantities of TiO_2 not to exceed 1% by weight of food. TiO_2 becomes transparent at the nanoscale (particle size < 100 nm), and is able to absorb and reflect UV light, making it useful in sunscreens. Nowadays, TiO_2 nanoparticles are used widely because of their high stability, anticorrosive character, and photocatalysis.

TiO_2 nanoparticles can produce free radicals with a strong oxidizing ability which can catalyze DNA damage both in vitro and in human cells [30]. TiO_2 nanoparticles also have pulmonary toxicity after endotracheal inhalation and instillation into the organism. It was reported that the TiO_2 nanoparticles (20 nm) penetrated more easily into the pulmonary interstitial area than the fine particles (250 nm) of the same mass [31]. The size-dependent toxicity of TiO_2 particles may not be significant if different routes of administration or different genders are used [32]. In an inhalation exposure study, mice exposed to 2–5 nm TiO_2 nanomaterials revealed a moderate inflammatory response among animals [33]. Pulmonary toxicity research in rats with three forms of TiO_2 particles showed that exposures to ultrafine TiO_2 particles can induce typical pulmonary effects, based on their composition and crystal structure [34].

4.2.2.3 Silica Nanomaterials

Nanomaterials of silica, a non-metal oxide, have been used in chemical mechanical polishing, and as an additive to drugs, cosmetics, printer toners, varnishes, and food, because it is a "generally regarded as safe" (GRAS) material. In recent years, applications of SiO_2 nanomaterials have been extended to biomedical and biotechnological fields, such as biosensors [35], biomarkers [36], cancer therapy [34, 37], DNA delivery [38], and enzyme immobilization [39].

Recent literature searches indicate that silica nanomaterials are nontoxic at low dosages but cell viability decreases at high dosages, because high dosages of silica induce cell membrane damage. On the other hand, silica-chitosan composite nanomaterials are known to induce less inhibition in cell proliferation and less membrane damage. The cytotoxicity of silica to human cells depends strongly on their metabolic activities, but it could be reduced by combining with chitosan [40]. In addition, dose-dependent exposure to SiO_2 nanoparticles induced cytotoxicity in human bronchoalveolar carcinoma-derived cells that was closely correlated to increased oxidative stress. It appears that SiO_2 nanomaterials reduce cell viability resulting from penetration of the particles into the cell nucleus [41].

4.3 Polymeric Carriers

Polymeric carriers are often used as drug delivery systems. They must not only be biocompatible and immunocompatible, but also be readily eliminated from the body, preferably through biodegradation. There are so many polymers that it is very difficult to classify each by certain criteria. For convenience, however, they are divided into biodegradable and non-biodegradable polymers in this chapter. The biodegradable polymers can be hydrophilic, hydrophobic, or amphiphilic.

4.3.1 Non-Biodegradable Polymers

Non-biodegradable polymers were frequently used as implant drug delivery systems in the early 1970s because of their long-lasting release and reduced host response. Examples are poly(vinyl alcohol) (PVA), poly(ethylene vinylacetate)

(PEVA) [42, 43], and polysulfone capillary fiber (PCF) [44]. They were proven to be safe in rabbit eyes for months [43, 44].

4.3.2 Biodegradable Polymers

4.3.2.1 Hydrophilic Polymers

Hydrophilic polymers generally have little immunogenicity in clinical applications. However, most hydrophilic macromolecules have to be crosslinked or copolymerized to form hydrogels; otherwise they will be dissolved and cleared from the body. Physical hydrogels may be better than the chemical ones, because most crosslinking agents are toxic and the chemical crosslinking process may chemically affect the entrapped molecules. It is very important to remove any residual crosslinking agent before in vivo application.

At the end of the last century, synthetic polymers became more and more important. Synthetic polymers could be tailored for various physicochemical properties to suit various applications. The synthetic polymers used in biomedical applications must be biocompatible, i.e., they must not provoke a defensive, potentially dangerous reaction in vivo. Application of nanomaterials may need to be considered as "polymer genomics." The term "polymer genomics" is defined as "an effect of synthetic polymers on pharmacogenomic responses to chemotherapeutic agents and the expression of transgenes delivered into cells" [45, 46]. Understanding of polymer genomics is expected to lead to safe and efficient nanoparticles for clinical applications.

Poly(2-hydroxypropyl methacrylamide) (PHPMA) is a potential water-soluble carrier. Rihova [47] reported that the molecular weight and the properties of the oligopeptidic side chains could result in some immunogenicity. PHPMA with molecular weight around 30 kDa is not recognized as a foreign molecule and has no recorded defense reaction to it. The attachment of pendant oligopeptide sequences to the HPMA backbone bestows a certain degree of immunogenicity, which depends on the composition of the oligopeptidic side chains, dose and route of application, molecular weight, and the genotype of the immunized individual.

4.3.2.2 Hydrophobic Polymers

Hydrophobic polymers are often used for long-term drug delivery, such as intraocular implants [48]. For convenience of avoiding removal of the system after completing drug release, biodegradable polymers are often preferred. Pure polyanhydrides can be degraded in 3 years unless it is copolymerized with sebacic acid (SA) [49]. Increasing the percent of the SA leads to faster degradation, and the copolymer with 80% SA can degrade in just a few days. The copolymer with SA is less hydrophobic and is known to be a good biocompatible material [50].

Poly(ortho ester) (POE) is a hydrophobic, biodegradable polymer. Currently, there are four (I–IV) families of POE [51, 52, 53]. There are acidic and basic portions in POE. The acidic portion determines the degradation rate. The basic portion maintains the polymer backbone's stability [54, 55]; moreover, it neutralizes the acidic microenvironment when the POE is degraded [56]. Different molecular weights of POE polymers have different release profiles in vitro [57]. Traditional gamma irradiation sterilization results in degradation of the POE III. So when a POE polymer is used, aseptic preparation is preferred [58]. During the storage of the POE III and its drug delivery system, they have to be sealed in glass bottles under an argon atmosphere [59].

Poly(lactic acid) (PLA), poly(glycolic acid) (PGA), and poly(lactic-co-glycolic acid) (PLGA) are the most widely used biodegradable polymers. The biggest problem of these polymers is that their hydrolysis results in acidic products, which induce protein degradation. In order to solve the problem, Jiang and Schwendeman [60, 61] used a blend of hydrophobic PLA and hydrophilic PEG for delivery of bovine serum albumin (BSA). With the PEG content in the blend, the degradation speed of PLA decreased, and BSA remained structurally intact without aggregation. PLGA/PEG blends showed the same effect [62]. There are other strategies, such as preparation through an o/o emulsion and co-encapsulation of additives [60, 61, 63], to alleviate the acidic microenvironment. The carboxyl end groups of the degradation products of the polymers can interact with positive charges of proteins to adversely affect the protein stability [64, 65].

During degradation of biodegradable polymers in the rabbit's eyes, triphasic release patterns were observed: initial drug-burst, diffusive phase, and a final drug burst. The latest procedure is generally uncontrollable and poorly predictable. Yasukawa [66, 67] reviewed the attempts to improve the release procedures. The larger the molecular weight or the lower the glycolide content, the slower the biodegradation [66]. Using PLA polymers with two different molecular weights in different ratios resulted in a decrease of the final drug burst with a pseudo-zero order kinetic of drug release [68].

4.3.2.3 Amphiphilic Polymers

Amphiphilic polymers are promising polymeric carriers as they can load the drug under a mild condition [69] as a polymer micelle or as a sol-gel phase reversible system. Amphiphilic polymers are divided into two parts: hydrophilic and hydrophobic segments. The properties of the two parts and the ratio of the two parts determine the in vivo fate of the micelles [70].

Poly(ethylene glycol) (PEG) [71, 72, 73], or poly(ethylene oxide) (PEO) [74], is the most commonly used polymer as the hydrophilic part of the polymer micelle. They endow the micelles a stealth surface by repelling the foreign substance, and thus increase the blood circulation half-life of the polymer micelles in vivo. It was reported that MPEG-PLA improved the efficacy of the direct nose–brain transport for drugs, which is especially important for peptides and proteins that are unable to penetrate through the blood-brain barrier [75].

A small difference between ethylene oxide (EO) and propylene oxide (PO) monomer units is the additional methyl group in the PO unit, which makes it more hydrophobic. The hydrophobic segment of the polymer containing PO units can be used to adsorb and anchor the block copolymer molecule to the nanoparticle surface, while the hydrophilic EO-containing segment, or PEG sections, can extend into solution and shield the surface of the particle. It was found, however, that the physically adsorbed polymers can be desorbed [76], and thus covalent grafting to the surface may be necessary for improved stability [77, 78]. Properties of PEG chains, such as molecular weight [72, 79, 80, 81], surface chain density [81, 82, 83], and conformation [72], affect biodistribution and pharmacokinetics of nanoparticles. The optimal molecular weight of the PEG chain for surface coating is above 2000; otherwise the length of the PEG chain is too short to be flexible [72, 81, 84].

An increased surface coverage by PEO, e.g., using Poloxamer 407, resulted in not only a reduction of the amount of adsorbed serum proteins, but also effected the type of proteins adsorbed [74]. When the surface coverage was above 25%, high-molecular-weight proteins did not adsorb onto the nanoparticles. Even at the 5% surface coverage, the in vivo circulation time was longer than the uncoated nanoparticle.

The nature and state of the hydrophobic segment in the micelles have a significant impact on the in vivo stability as well as the pharmacokinetics and biodistribution of the micelles. The longer length of the hydrophobic segment and the higher proportion of the hydrophobic polymer endow a greater thermodynamic stability [85, 86]. The physical state of the core-forming polymer, such as amorphous, crystalline or semicrystalline, has an effect on the micelle stability [87, 88]. A micelle containing a hydrophobic block with a glass transition temperature (T_g) exceeding 37°C has a "frozen" core. The crystalline or semi-crystalline [88] core makes the micelle more stable and the duration of drug release becomes longer.

The ratio of the hydrophilic/hydrophobic will also affect the shape of the micelle. In general, when the hydrophobic part of the micelle overweighs the hydrophilic part [87] or the length of the hydrophilic segment is longer than that of the core block [89], the shape of the resulting micelles is spherical. On the other hand, non-spherical structures, including rods and lamellae, can be formed by increasing the length of the core segment beyond that of the corona-forming chains or the hydrophilic part of the micelle overweighs the hydrophobic part.

Amphiphilic β-cyclodextrin nanosphere suspensions [90] also reported to have unexpected good physical stability of the suspensions after 3 years of storage at room temperature after the secondary hydroxyl functions of the β-CDa glucopyranosyl units were grafted by hexanoyl carbon chains.

4.3.3 Others

The copolymer concentration is another factor affecting the effectiveness of the carrier. Three MPEG-b-poly(ε-caprolactone) (MPEG-b-PCL) copolymer concentrations, 0.2 mg/kg dose group (unimers), 2 mg/kg dose group (unstable micelle), and the 250 mg/kg dose group (stable micelle), were used to investigate the in vivo fate of mice following intravenous administration [88]. It was found that when the polymer was given in unimers, more copolymer penetrated into tissues. The thermodynamically unstable micelles (i.e., 2 mg/kg dose group) had a much longer circulation half-life and slower rate of elimination than the unimers. The thermodynamically stable micelles (i.e., 250 mg/kg dose group) had the slowest elimination rate from the plasma during the elimination phase. The 250 mg/kg dose group also had the lowest tissue to plasma concentration ratio during the elimination phase.

Copolymers with a low polydispersity index (PDI) (e.g., PDI < 1.1) usually lead to more stable micelle systems in vivo [88]. The polymer purity and molecular weight distribution are also known to affect microsphere morphology and in vitro cytotoxicity [91]. Microspheres showed decreased in vivo degradation rate and lower initial protein burst after ultrafiltration. Ultrafiltration appears to be a useful method to control the properties of microspheres.

4.4 Additives

Pharmaceutical additives are widely used to preserve the pharmacological activity of drugs and to prolong the shelf life of the dosage forms, especially of protein drugs. Stabilization of protein drugs by additives is based on the surface-active properties of some additives and/or electrostatic interactions. However, there is no general rule how to choose an optimal additive for a specific protein.

Sugars have been widely used as excipients for stabilizing protein drugs. No general rules have been established explaining how sugars stabilize proteins. Different sugars have different effects on the same protein, whereas the same sugar also has different effects on different proteins. For example, trehalose and mannitol have significant protective effects on the soluble non-aggregated interferon- γ (INF-γ) and growth hormone after emulsification and ultrasonication [16], whereas no or little protecting effect on insulin-like growth factor-I [92]. Trehalose, mannitol, and sucrose have no protecting effect against the degradation of lysozyme, whereas lactose and lactulose have signification positive effects [93, 94]. Cyclodextrins (CD) [95, 96] is also used as a special stabilizer. Hydroxypropyl-β-cyclodextrin is known to increase the stability of recombinant human INF-α-2a protein [97]. Generally, sugars are added in the inner aqueous phase in the w/o/w emulsion process.

A surfactant is another additive which is widely used in protein formulations. Non-ionic surfactants usually have better effects than ionic surfactants because binding of ionic surfactants to proteins can cause protein denaturation. Poloxamer 188 successfully prolonged the release of active INF-α when it was mixed with PLGA [98], whereas it had no effect on BSA secondary structure [17].

Albumins and gelatins are frequently added to the inner aqueous phase during emulsion process to protect the bioactivity of proteins. The protective function of albumins (i.e., bovine, human or rat serum albumins) results from their surface-active properties. Albumins occupy the interfaces to shield the pharmacological protein from exposure to solvent [92, 99, 100]. Albumins are also known to remove the protons formed during degradation of PLGA, avoiding protein aggregation resulting from the acidic environment [101]. Albumins, however, may not work in many other cases [102]. Gelatin is another protein which is used as an additive to protect protein drugs from ultrasonication [92]. The gelatin protective effect is dependent on the gelatin molecular weight and concentration; the higher the molecular weight/concentration, the better the stabilization effect [103].

Some synthetic polymers are also used as an additive too. PEG has been used to protect protein against degradation [102, 103, 104], although sometimes it resulted in adverse affects [105]. PEG can be added to either aqueous or organic phases [106, 107]. Two or more additives can be combined to enhance the protective function over individual additives [92, 108].

4.5 Structure

4.5.1 Quantum Dots

Quantum dots (QDs) are semiconductor nanocrystals with unique optical and electrical properties. QDs have a longer durability and higher fluorescence than conventional organic fluorophores, thus it can act as information and visual technologies to transfer in vivo imaging and diagnostics of living organisms [109]. Moreover, fluorescent QDs can be joined together with bioactive moieties (e.g., antibodies, receptor ligands) to target specific biological event and cellular structures and receptors.

To understand the potential toxicity of QDs, it is required to understand the physicochemical properties of QDs. QDs consist of a metalloid crystalline core and a shell that shields the core and makes the QD bioactive. Many QD core metals (e.g., Cd, Pb, Se) are known to be toxic to vertebrate systems even at low concentrations (parts per million). For example, Cd, a bioaccumulative carcinogen, has a biologic half-life of 15–20 years in humans, can cross the blood-brain barrier and placenta, and is distributed throughout the body. Degradation of the QD coating may also result in unexpected reactions of the QD in vivo. Furthermore, some QD coating materials, such as mercaptoacetic acid, have been found to be cytotoxic. Till now, there are no standardized protocols on QD synthesis and coating. The safety of QDs is known to depend on QD size, charge, concentration, and bioactivity of outer coating (capping material and functional groups) [110, 111]. Protonation [112] or photo-oxidation [113] is known to deteriorate the stability of QDs. In a relatively low pH range, between 2 and 7, the deprotonated thiols or thiolates which are bound to cadmium chalcogenide QDs will become protonated and detached from the QD surface causing the precipitation of the crystals [112]. Photooxidation of the surface ligands also makes them detached from the surface, leading to precipitation. Increasing the thickness and packing density of the ligand is useful to delay the initiation process of photooxidation [113].

4.5.2 Carbon Nanotubes

Carbon nanotubes (CNTs) consist exclusively of carbon atoms arranged in a series of condensed benzene rings rolled-up into a tubular structure. Various physico-chemical properties of CNTs, such as an ordered structure with high aspect ratios, ultralight weight, high mechanical strength, high electrical conductivity, high thermal conductivity, metallic or semi-metallic behavior and high surface area, present unique opportunities for diverse applications [114, 115].

Non-functionalized CNTs are hydrophobic materials, requiring functionalization to be compatible in the biological media. Such functional-groups can be obtained by adsorption, electrostatic interaction, or covalent bonding of different molecules to make CNTs more hydrophilic [116]. The physical state of the CNTs is also very important to the safety of the usage of the CNTs [117, 118]. The functionalized surface can prevent aggregation of the individual tubes that occurs through van der Waals forces.

4.5.3 Dendrimers

Dendrimers consist of a central core molecule acting as the root from which a number of highly branched, tree-like arms originate in an ordered and symmetric fashion [119]. Because of their unique molecular architecture, they have many distinctive properties which are different from other polymers, such as the gradual stepwise synthesis method through a divergent or a convergent one, a well-defined size and structure with a low PDI. The relatively empty intermolecular cavity and the highly dense terminal groups can be used to entrap host molecules.

Most dendrimers are known to have poor solubility in aqueous solutions, and a structure that would predict a likelihood of cellular accumulation. They may present unacceptable toxicity and/or immunogenicity if administered parenterally. It is well known that the large surface area/volume of all nanosized materials can potentially lead to unfavorable biological responses if they are inhaled and absorbed into the body [7, 120]. It is widely known that dendrimers with $-NH_2$ termini display concentration- and generation-dependent cytotoxicity [121]. The exposure time is also an important factor which affects the morphology of the cell. When polyamidoamine (PAMAM) with generation 4 was incubated in B16F10 cells for 5 hrs, cell membranes were damaged [122], but it was reversible by removing the dendrimers [123].

Dendrimer cytotoxity is not only dependent on the chemistry of the core, but is also influenced by the properties of the surface. PAMAM dendrimer and polypropylenimine dendrimer have a cationic net surface charge. Cationic surface charges are in general more toxic than anionic or PEGylated dendrimers [124]. In order to decrease the toxicity, quaternization is always used as a strategy [125, 126]. Increased branching (or generation) and a greater surface coverage with biocompatible terminal groups, such as C12 lauroyl groups or PEG 2000, reduce the dendrimer toxicity significantly [122]. However, when the surface of generation 4 PAMAM dendrimers was modified with lysine or arginine, the toxicity increased which was confirmed by incubation with HepG2 or 293T human embryonic kidney cells for 48 hrs, probably due to the increased density and molecular weight [127]. To improve the transfection efficiency, some compounds are added as additives into the dendrimer–DNA complexes, such as DEAE-dextran [128] and substituted cyclodextrin [129].

4.6 Processes

The particle size plays a key role in the final biodistribution and pharmacokinetics. It was reported [84] that a particle with hydrodynamic radius of over 200 nm will be cleared more quickly than particles with radius under 200 nm. Controlling the particle size may be one way to prepare safe nano/micro materials.

4.6.1 Emulsion

Emulsion methods have been used widely in preparing nano/micro particles. A Water-in-oil-in-water (w/o/w) double emulsion method has been most widely used. Many hydrophilic drugs/proteins can be encapsulated by this method. The particles obtained in this method are very stable [130]. In emulsion methods, protein drugs can become denatured by exposure to the interface between water and solvent [99, 131]. The effect of the primary w/o emulsion has a stronger denaturing effect than that of the secondary w/o/w emulsion [132]. To avoid the exposure of proteins to organic solvent, protein particles in the solid state can be directly suspended in the organic phase to form s/o/w emulsion [23, 133]. For this s/o/w method to work, the hydrophilic drug/protein power has to be under an anhydrous condition before encapsulation [17, 18]. The anhydrous proteins can be obtained by freeze-drying or SFD before being encapsulated (See Section 2). The nature of the organic solvent also has an impact on protein stability. Use of blend solvents (such as acetone/ethylene chloride blend) could reduce surface tension between the organic and the water phases [20, 23, 102].

Although all ultrasound, sonication, vortexing, and homogenization operations can result in protein degradation, a good choice of the apparatus can minimize the instability of proteins [134]. Milling is a traditional method for micronization of drug powders. The standard micronization processes comprise crushing/milling, air micronization, sublimation, and recrystallization from solvents. There are various mills, such as ball mills, colloid mills, and jet or fluid energy mills. Most of them have advantages and shortcomings. The mechanical treatments can damage and degrade particles due to high stresses (thermal and mechanical) generated by attrition. They result in particle adhesion, agglomeration, and loss of drug activity. Jet milling is a process to reduce the size of crystals or coarse particles by high velocity air [135]. The majority of inhalation powders are prepared by jet mill [135, 136]. It can produce particles between 1 and 20 μm [136]. In addition to milling, crystallization, and lyophilization have been also widely used. For crystallization processes, co-precipitating solvent is often used [137]. Some solvents will result in the instability of the biopharmaceuticals, but the use of optimal lyoprotectant can prevent aggregation and increase shelf life [138, 139].

4.6.2 Spray Drying

Spray drying transforms protein-containing solutions to powder in a single step [140]. The major advantage of the process is that this technology can avoid thermal degradation. Though high temperature drying air is used, the drying time usually lasts less than 100 milliseconds to seconds [135, 136]. During the drying process, the material temperature remains significantly less than the drying air due to evaporative cooling.

There are some problems to consider with spray drying. Atomization may result in degradation and denaturation of proteins. This problem has been alleviated by adding suitable excipients, such as sucrose, trehalose, lactose, PVA, dipamitoylphosphatidylcholine, and even albumin [141, 142, 143, 144, 145, 146, 147]. The yield of spray drying is rather low in the range of 20–50% [136, 146]. This can be improved to 70% by introduction of high-performance cyclone for collecting the dried particles [135, 148]. It is difficult to control the mean droplet size during spray drying [149]. Though the use of ultrasonic nozzles can lighten the problem, it can also cause protein denaturation [150]. The spray drying technique is difficult with poorly water-soluble drugs. In that case, spray freezing into liquid (SFL) can be used. Careful selecting the operating parameters can play a significant role in obtaining high quality product in spray drying.

4.6.3 Spray Freeze-Drying

Spray freeze-drying is a process which takes advantage of the very low boiling point of nitrogen, oxygen, or argon to freeze a solution containing proteins, then to lyophilize the frozen droplets to obtain porous spherical particles. In contrast to the dense particles (∼3 μm) produced by spray drying, SFD results in porous, fragile particles (∼8-10 μm) with low aerodynamic size [151].

There are a few limitations of SFD. The process is time-consuming (taking 3 days) and expensive because of the safety issues resulting from the extremely low boiling point (below –195.8°C) of liquid nitrogen. The stress associated with freezing and drying, especially the adsorption of a protein at the air–liquid interface during atomization, results in irreversible damage to the protein. This problem can be reduced by limiting the time of exposure to the air–liquid interface during atomization [152]. SFL is an improved SFD technique.

Spray freezing into liquid [153, 154, 155, 156] has been developed for poorly water soluble or insoluble drugs. The advantage of the technique is that the aqueous protein solution is sprayed directly into the liquid nitrogen through an insulated nozzle instead of into the cold vapor in SFD. Another improved technique is spray-freezing with compressed CO_2 [135]. In this technique, biopharmaceutical and excipients mixture is atomized, then mixed with compressed CO_2, developing a CO_2-saturated aqueous solution. The droplets become frozen particles when they are sprayed through a nozzle of a sprayer. The goal of the method is to obtain stable, porous or hollow protein particles with a narrow size distribution [157].

4.6.4 Supercritical Fluid (SF) Technology

Supercritical fluid technique [135, 158] combines advantages of both liquid and gas. The density values and solvation power of a solute can be adjusted by the SF's critical temperature and pressure. The viscosity of the solutes in SF is lower than liquid, while the diffusivity is higher. Most important is the SFs are highly compressible. The most widely used SF is CO_2 because of its low critical temperature (31.2°C) and pressure (7.4 MPa), non-flammable, non-toxic, and inexpensive. The only limitation of SF CO_2 is its limited solvation power though it can be changed into an advantage when the SF CO_2 is used as an anti-solvent. The solvation power of a SF CO_2 can be adjusted by incorporating a small amount of volatile cosolvent, such as acetone or ethanol, which acts as organic modifier.

In general, SFs can be divided into 3 groups.

- Precipitation from supercritical solutions composed of SF and solutions (rapid expansion of supercritical solution, RESS).
- Precipitation from gas saturated solutions (precipitation from gas-saturated solution, PGSS).
- Precipitation from saturated solutions using SF as anti-solvent (including gaseous anti-solvent, GAS; aerosol solvent extraction system, ASES; solution enhanced dispersion by SF, SEDS and precipitation by compressed anti-solvent, PCA) [159].

Recently innovative techniques have been developed where the SF CO_2 is used to assist spray drying. The SF and the solution are intimately mixed and sprayed in a drying atmosphere. This process allows the minimal decomposition of thermally labile drugs, no need of high pressure vessel and as small as 3 μm particles.

4.7 Workers' Safety

Although more than 35 countries have developed various R&D programs on nanotechnology since 2000, the importance of safety issues regarding nanoparticles was recognized only recently. In 2006, the National Institute of Occupational Safety and Health (NIOSH) published a document, "Approaches to safe nanotechnology." It reviews potential risks of nanotechnology at workplace in order for workers, employers, researchers, and the general public to be aware of the hazard of nano-materials and to minimize exposure to nanostructures [160].

The way workers are exposed to engineered nanoparticles is directly related to safety and health. The exposure routes can be inhalation, dermal, and ingestion, which were reviewed already [8, 161, 162, 163]. Pathophysiology and toxicity of nanoparticles in the body include reactive oxygen species generation, oxidative stress, mitochondrial perturbation, inflammation, uptake by reticulo-endothelial system, protein degradation/denaturation, brain/peripheral nervous system injury, DNA damage, endothelial dysfunction/blood clotting, and alternation of cell cycle [8, 162, 164, 165]. Several possible mechanisms on nanoparticle–biological tissue interactions have been suggested. They include UV activation leading to radical production, impurities and defects to induce active electronic configuration, redox cycling and catalytic activity of surface metals and polymers, and particle dissolution in media [165].

Although the exact mechanism of nanoparticle toxicity is not understood very well, several key factors have been suggested, which are size, surface area/chemistry, solubility, and shape [164]. It was reported that smaller particles were more penetrative into lung tissue than larger ones [166]. In addition, ultra-fine particles (UFP, < 100 nm) rather than fine (< 2.5 μm) or coarse particles (2.5–10 μm) could penetrate into even cells and be localized at mitochondria leading to oxidative stress [167]. Surface area, which is related to the dose of nanoparticles, is exponentially related to lung deposition of nanoparticle, tissue damage, and inflammation [164]. Insoluble nanoparticles were known to be retained in lung tissues and induce inflammation depending on the surface area, which was initiated by oxidative stress [168]. Special interest on the shape was initiated from single-walled nanotube (SWNT), which has 0.7–1.5 nm in diameter and several micrometers in length. In vitro incubation of SWNT with keratinocyte induced oxidative stress [169] and inflammation [170]. Toxicological aspects of nanoparticles should be considered before developing, preparing, and applying nanoparticles to avoid unexpected hazardous conditions of workers and patients.

4.8 The Related Guidance

Because of extremely fast advances in nanotechnology, making appropriate guideline on the biocompatibility of nanoparticles in clinical applications seems relatively very slow. A brief review of existing guidance would be beneficial for development of better and safer nanoparticles.

(1) The International Standards Organization 10993 (ISO 10993/FDA #G95-1/Japanese guideline

ISO 10993 consists of 20 parts of harmonized standards for biocompatibility since 1986. However, the ISO 10993 has too many details and it is not free to access so that it may be considered only as a suggestion rather than a standard on biocompatibility of nanoparticles. The FDA published blue book memorandum #G95-1 entitled "Required biocompatibility training and toxicology profiles for evaluation of medical devices" in 1995. It lists a brief and broad guideline for biocompatibility tests properly based on ISO 10993, which includes cytotoxicity, sensitization, irritation or intracutaneous reactivity, acute system toxicity, sub-chronic toxicity, genotoxicity, implantation, and hemocompatibility. The guideline also includes detailed categories, such as contact time of engineered materials to host tissue and device types to be applied. Differences among the international Standards Organization 10993 (ISO 10993), FDA #G95-1 and Japanese Guideline were compared [171].

(2) A guideline about the critical path to medical device development [172]

The summary of this guideline published by the FDA in 2004 is shown in Fig. 4.1. It shows the pathway to modify the nanoparticles for clinical applications. Three important points to be considered are safety, medical utility, and industrialization. Engineered nanoparticles should be safe enough to be applied to humans. The medical utility means efficacy of the developed nanoparticles for the benefit of human health. Moreover, if nanoparticles cannot be produced in large scale, it will be less useful for the public health. At each stage of development, obtained results and data

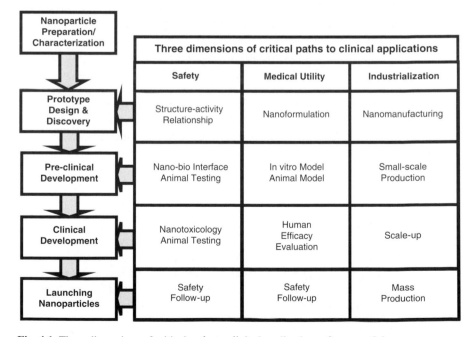

Fig. 4.1 Three dimensions of critical paths to clinical applications of nanoparticles

should be examined in terms of these three factors. The guideline did not present any detailed protocols, but general test methods were described in the FDA memorandum No. G95-1.

(3) FDA guideline for industry and FDA staff [173]

Although appropriate guidelines and protocols for clinical applications of nanoparticles are urgently needed, the difficulty is in that nanoparticles can not only be used by themselves, but also be combined with different devices. Thus, a combination product defined by FDA in this guideline could be "a product comprised any combination of a drug and a device, a biological product and a device, a drug and a biological product, or a drug, device, and a biological product." The report describes that combination products may require more careful consideration during development than conventional products due to their complexity. Such complexity may not only be simply due to combination of devices, drugs, and biological products, but also due to interactions between a combination product (as well as its constituents) and biological tissues.

(4) FDA Guidance for industry, investigators, and reviewers [174] and (5) FDA final guidance for industry and CDRH staff [175]

For clinical applications, additional guidelines were suggested in the report, which described general approaches for investigating new drugs and for modification of existing devices or protocols.

References

1. Jones R (2004) The future of nanotechnology. Phys World 17:25–29
2. Gogotsi Y (2003) How safe are nanotubes and other nanofilaments? Mater Res Innov 7: 192–194
3. Hoet PH, Bruske-Hohlfeld I, Salata OV (2004) Nanoparticles – known and unknown health risks. J Nanobiotechnology 2:12
4. Robichaud CO, Tanzil D, Weilenmann U, Wiesner MR (2005) Relative risk analysis of several manufactured nanomaterials: an insurance industry context. Environ Sci Technol 39: 8985–8994
5. IRGC (2006) Survey on nanotechnology governance. International Risk Governance Council, Geneva, Switzerland
6. Munshia D, Kurianb P, Bartlettc RV, Lakhtakia A (2007) A map of the nanoworld: sizing up the science, politics, and business of the infinitesima. Futures 39:432–452
7. Giles J (2004) Size matters when it comes to safety, report warns. Nature 430:599
8. Maynard A (2006) Nanotechnology: assessing the risks. Nanotoday 1:22–33
9. Papageorgiou I, Brown C, Schins R, Singh S, Newson R, Davis S, Fisher J, Ingham E, Case CP (2007) The effect of nano- and micron-sized particles of cobalt-chromium alloy on human fibroblasts in vitro. Biomaterials 28:2946–2958
10. Bilati U, Allemann E, Doelker E (2005) Strategic approaches for overcoming peptide and protein instability within biodegradable nano- and microparticles. Eur J Pharm Biopharm 59:375–388
11. Harris JM, Chess RB (2003) Effect of pegylation on pharmaceuticals. Nat Rev Drug Discov 2:214–221

12. Diwan M, Park TG (2001) Pegylation enhances protein stability during encapsulation in PLGA microspheres. J Control Release 73:233–244
13. Diwan M, Park TG (2003) Stabilization of recombinant interferon-alpha by pegylation for encapsulation in PLGA microspheres. Int J Pharm 252:111–122
14. Fernandes AI, Gregoriadis G (2001) The effect of polysialylation on the immunogenicity and antigenicity of asparaginase: implication in its pharmacokinetics. Int J Pharm 217: 215–224
15. Gregoriadis G, Fernandes A, Mital M, McCormack B (2000) Polysialic acids: potential in improving the stability and pharmacokinetics of proteins and other therapeutics. Cell Mol Life Sci 57:1964–1969
16. Cleland JL, Jones AJ (1996) Stable formulations of recombinant human growth hormone and interferon-gamma for microencapsulation in biodegradable microspheres. Pharm Res 13:1464–1475
17. Carrasquillo KG, Stanley AM, Aponte-Carro JC, De Jesus P, Costantino HR, Bosques CJ, Griebenow K (2001) Non-aqueous encapsulation of excipient-stabilized spray-freeze dried BSA into poly(lactide-co-glycolide) microspheres results in release of native protein. J Control Release 76:199–208
18. Carrasquillo KG, Carro JC, Alejandro A, Toro DD, Griebenow K (2001) Reduction of structural perturbations in bovine serum albumin by non-aqueous microencapsulation. J Pharm Pharmacol 53:115–120
19. Stevenson CL (2000) Characterization of protein and peptide stability and solubility in non-aqueous solvents. Curr Pharm Biotechnol 1:165–182
20. Stevenson CL, Tan MM (2000) Solution stability of salmon calcitonin at high concentration for delivery in an implantable system. J Pept Res 55:129–139
21. Yoo HS, Choi HK, Park TG (2001) Protein-fatty acid complex for enhanced loading and stability within biodegradable nanoparticles. J Pharm Sci 90:194–201
22. Huyghues-Despointes BM, Qu X, Tsai J, Scholtz JM (2006) Terminal ion pairs stabilize the second beta-hairpin of the B1 domain of protein G. Proteins 63:1005–1017
23. Bilati U, Allemann E, Doelker E (2005) Nanoprecipitation versus emulsion-based techniques for the encapsulation of proteins into biodegradable nanoparticles and process-related stability issues. AAPS PharmSciTech 6:E594–604
24. Jen A, Merkle HP (2001) Diamonds in the rough: Protein crystals from a formulation perspective. Pharm Res 18:1483–1488
25. Elkordy AA, Forbes RT, Barry BW (2002) Integrity of crystalline lysozyme exceeds that of a spray-dried form. Int J Pharm 247:79–90
26. Elkordy AA, Forbes RT, Barry BW (2004) Stability of crystallised and spray-dried lysozyme. Int J Pharm 278:209–219
27. Lee MJ, Kwon JH, Shin JS, Kim CW (2005) Microcrystallization of alpha-lactalbumin. J Cryst Growth 282:434–437
28. Berry CC, Curtis ASG (2003) Functionalisation of magnetic nanoparticles for applications in biomedicine. J Phys D Appl Phys 36:R198–R206
29. Pisanic TR, 2nd, Blackwell JD, Shubayev VI, Finones RR, Jin S (2007) Nanotoxicity of iron oxide nanoparticle internalization in growing neurons. Biomaterials 28:2572–2581
30. Dunford R, Salinaro A, Cai L, Serpone N, Horikoshi S, Hidaka H, Knowland J (1997) Chemical oxidation and DNA damage catalysed by inorganic sunscreen ingredients. FEBS Lett 418:87–90
31. Ferin J, Oberdorster G, Penney DP (1992) Pulmonary retention of ultrafine and fine particles in rats. Am J Respir Cell Mol Biol 6:535–542
32. Wang J, Zhou G, Chen C, Yu H, Wang T, Ma Y, Jia G, Gao Y, Li B, Sun J, Li Y, Jiao F, Zhao Y, Chai Z (2007) Acute toxicity and biodistribution of different sized titanium dioxide particles in mice after oral administration. Toxicol Lett 168:176–185
33. Grassian VH, O'Shaughnessy PT, Adamcakova-Dodd A, Pettibone JM, Thorne PS (2007) Inhalation exposure study of titanium dioxide nanoparticles with a primary particle size of 2 to 5 nm. Environ Health Perspect 115:397–402

34. Warheit DB, Webb TR, Reed KL, Frerichs S, Sayes CM (2007) Pulmonary toxicity study in rats with three forms of ultrafine-TiO2 particles: differential responses related to surface properties. Toxicology 230:90–104

35. Zhang FF, Wan Q, Li CX, Wang XL, Zhu ZQ, Xian YZ, Jin LT, Yamamoto K (2004) Simultaneous assay of glucose, lactate, L-glutamate and hypoxanthine levels in a rat striatum using enzyme electrodes based on neutral red-doped silica nanoparticles. Anal Bioanal Chem 380:637–642

36. Santra S, Zhang P, Wang K, Tapec R, Tan W (2001) Conjugation of biomolecules with luminophore-doped silica nanoparticles for photostable biomarkers. Anal Chem 73: 4988–4993

37. Hirsch LR, Stafford RJ, Bankson JA, Sershen SR, Rivera B, Price RE, Hazle JD, Halas NJ, West JL (2003) Nanoshell-mediated near-infrared thermal therapy of tumors under magnetic resonance guidance. Proc Natl Acad Sci USA 100:13549–13554

38. Bharali DJ, Klejbor I, Stachowiak EK, Dutta P, Roy I, Kaur N, Bergey EJ, Prasad PN, Stachowiak MK (2005) Organically modified silica nanoparticles: a nonviral vector for in vivo gene delivery and expression in the brain. Proc Natl Acad Sci USA 102:11539–11544

39. Qhobosheane M, Santra S, Zhang P, Tan W (2001) Biochemically functionalized silica nanoparticles. Analyst 126:1274–1278

40. Chang JS, Chang KL, Hwang DF, Kong ZL (2007) In vitro cytotoxicitiy of silica nanoparticles at high concentrations strongly depends on the metabolic activity type of the cell line. Environ Sci Technol 41:2064–2068

41. Lin W, Huang YW, Zhou XD, Ma Y (2006) In vitro toxicity of silica nanoparticles in human lung cancer cells. Toxicol Appl Pharmacol 217:252–259

42. Smith TJ, Pearson PA, Blandford DL, Brown JD, Goins KA, Hollins JL, Schmeisser ET, Glavinos P, Baldwin LB, Ashton P (1992) Intravitreal sustained-release ganciclovir. Arch Ophthalmol 110:255–258

43. Driot JY, Novack GD, Rittenhouse KD, Milazzo C, Pearson PA (2004) Ocular pharmacokinetics of fluocinolone acetonide after Retisert intravitreal implantation in rabbits over a 1-year period. J Ocul Pharmacol Ther 20:269–275

44. Rahimy MH, Peyman GA, Chin SY, Golshani R, Aras C, Borhani H, Thompson H (1994) Polysulfone capillary fiber for intraocular drug delivery: in vitro and in vivo evaluations. J Drug Target 2:289–298

45. Kabanov AV (2006) Polymer genomics: an insight into pharmacology and toxicology of nanomedicines. Adv Drug Deliv Rev 58:1597–1621

46. Kabanov AV, Batrakova EV, Sriadibhatla S, Yang Z, Kelly DL, Alakov VY (2005) Polymer genomics: shifting the gene and drug delivery paradigms. J Control Release 101:259–271

47. Rihova B (2007) Biocompatibility and immunocompatibility of water-soluble polymers based on HPMA. Compos Part B-Eng 38:386–397

48. Bourges JL, Bloquel C, Thomas A, Froussart F, Bochot A, Azan F, Gurny R, BenEzra D, Behar-Cohen F (2006) Intraocular implants for extended drug delivery: therapeutic applications. Adv Drug Deliv Rev 58:1182–1202

49. Leong KW, Brott BC, Langer R (1985) Bioerodible polyanhydrides as drug-carrier matrices. I: characterization, degradation, and release characteristics. J Biomed Mater Res 19: 941–955

50. Leong KW, D'Amore PD, Marletta M, Langer R (1986) Bioerodible polyanhydrides as drug-carrier matrices. II. Biocompatibility and chemical reactivity. J Biomed Mater Res 20:51–64

51. Heller J (2005) Ocular delivery using poly(ortho esters). Adv Drug Deliv Rev 57:2053–2062

52. Heller J, Barr J, Ng SY, Shen HR, Schwach-Abdellaoui K, Einmahl S, Rothen-Weinhold A, Gurny R (2000) Poly(ortho esters) – their development and some recent applications. Eur J Pharm Biopharm 50:121–128

53. Einmahl S, Capancioni S, Schwach-Abdellaoui K, Moeller M, Behar-Cohen F, Gurny R (2001) Therapeutic applications of viscous and injectable poly(ortho esters). Adv Drug Deliv Rev 53:45–73

54. Zignani M, Einmahl S, Baeyens V, Varesio E, Veuthey JL, Anderson J, Heller J, Tabatabay C, Gurny R (2000) A poly(ortho ester) designed for combined ocular delivery of dexamethasone sodium phosphate and 5-fluorouracil: subconjunctival tolerance and in vitro release. Eur J Pharm Biopharm 50:251–255
55. Einmahl S, Zignani M, Varesio E, Heller J, Veuthey JL, Tabatabay C, Gurny R (1999) Concomitant and controlled release of dexamethasone and 5-fluorouracil from poly(ortho ester). Int J Pharm 185:189–198
56. Zignani M, Le Minh T, Einmahl S, Tabatabay C, Heller J, Anderson JM, Gurny R (2000) Improved biocompatibility of a viscous bioerodible poly(ortho ester) by controlling the environmental pH during degradation. Biomaterials 21:1773–1778
57. Merkli A, Heller J, Tabatabay C, Gurny R (1994) Semi-solid hydrophobic bioerodible poly(ortho ester) for potential application in glaucoma filtering surgery. J Control Release 29:105–112
58. Sintzel MB, Schwach-Abdellaoui K, Mader K, Stosser R, Heller J, Tabatabay C, Gurny R (1998) Influence of irradiation sterilization on a semi-solid poly(ortho ester). Int J Pharm 175:165–176
59. Merkli A, Heller J, Tabatabay C, Gurny R (1996) Purity and stability assessment of a semi-solid poly(ortho ester) used in drug delivery systems. Biomaterials 17:897–902
60. Jiang W, Schwendeman SP (2001) Stabilization of a model formalinized protein antigen encapsulated in poly(lactide-co-glycolide)-based microspheres. J Pharm Sci 90: 1558–1569
61. Jiang W, Schwendeman SP (2001) Stabilization and controlled release of bovine serum albumin encapsulated in poly(D, L-lactide) and poly(ethylene glycol) microsphere blends. Pharm Res 18:878–885
62. Yeh MK (2000) The stability of insulin in biodegradable microparticles based on blends of lactide polymers and polyethylene glycol. J Microencapsul 17:743–756
63. Elvassore N, Bertucco A, Caliceti P (2001) Production of insulin-loaded poly(ethylene glycol)/poly(l-lactide) (PEG/PLA) nanoparticles by gas antisolvent techniques. J Pharm Sci 90:1628–1636
64. Cleland JL, Duenas ET, Park A, Daugherty A, Kahn J, Kowalski J, Cuthbertson A (2001) Development of poly-(D,L-lactide–coglycolide) microsphere formulations containing recombinant human vascular endothelial growth factor to promote local angiogenesis. J Control Release 72:13–24
65. Kostanski JW, Thanoo BC, DeLuca PP (2000) Preparation, characterization, and in vitro evaluation of 1- and 4-month controlled release orntide PLA and PLGA microspheres. Pharm Dev Technol 5:585–596
66. Yasukawa T, Kimura H, Tabata Y, Ogura Y (2001) Biodegradable scleral plugs for vitreoretinal drug delivery. Adv Drug Deliv Rev 52:25–36
67. Yasukawa T, Ogura Y, Sakurai E, Tabata Y, Kimura H (2005) Intraocular sustained drug delivery using implantable polymeric devices. Adv Drug Deliv Rev 57:2033–2046
68. Kunou N, Ogura Y, Yasukawa T, Kimura H, Miyamoto H, Honda Y, Ikada Y (2000) Long-term sustained release of ganciclovir from biodegradable scleral implant for the treatment of cytomegalovirus retinitis. J Control Release 68:263–271
69. Lee SH, Zhang ZP, Feng SS (2007) Nanoparticles of poly(lactide) – Tocopheryl polyethylene glycol succinate (PLA-TPGS) copolymers for protein drug delivery. Biomaterials 28: 2041–2050
70. Maysinger D, Lovric J, Eisenberg A, Savic R (2007) Fate of micelles and quantum dots in cells. Eur J Pharm Biopharm 65:270–281
71. Otsuka H, Nagasaki Y, Kataoka K (2003) PEGylated nanoparticles for biological and pharmaceutical applications. Adv Drug Deliv Rev 55:403–419
72. Peracchia MT, Vauthier C, Passirani C, Couvreur P, Labarre D (1997) Complement consumption by poly(ethylene glycol) in different conformations chemically coupled to poly(isobutyl 2-cyanoacrylate) nanoparticles. Life Sci 61:749–761

73. Moghimi SM (1997) Prolonging the circulation time and modifying the body distribution of intravenously injected polystyrene nanospheres by prior intravenous administration of poloxamine-908. A 'hepatic-blockade' event or manipulation of nanosphere surface in vivo? Biochim Biophys Acta 1336:1–6

74. Stolnik S, Daudali B, Arien A, Whetstone J, Heald CR, Garnett MC, Davis SS, Illum L (2001) The effect of surface coverage and conformation of poly(ethylene oxide) (PEO) chains of poloxamer 407 on the biological fate of model colloidal drug carriers. Biochim Biophys Acta 1514:261–279

75. Zhang QZ, Zha LS, Zhang Y, Jiang WM, Lu W, Shi ZQ, Jiang XG, Fu SK (2006) The brain targeting efficiency following nasally applied MPEG-PLA nanoparticles in rats. J Drug Target 14:281–290

76. Neal JC, Stolnik S, Schacht E, Kenawy ER, Garnett MC, Davis SS, Illum L (1998) In vitro displacement by rat serum of adsorbed radiolabeled poloxamer and poloxamine copolymers from model and biodegradable nanospheres. J Pharm Sci 87:1242–1248

77. Harper GR, Davies MC, Davis SS, Tadros TF, Taylor DC, Irving MP, Waters JA (1991) Steric stabilization of microspheres with grafted polyethylene oxide reduces phagocytosis by rat Kupffer cells in vitro. Biomaterials 12:695–700

78. Bazile D, Prudhomme C, Bassoullet MT, Marlard M, Spenlehauer G, Veillard M (1995) Stealth Me.Peg-Pla nanoparticles avoid uptake by the mononuclear phagocytes system. J Pharm Sci 84:493–498

79. Neuzillet Y, Giraud S, Lagorce L, Eugene M, Debre P, Richard F, Barrou B (2006) Effects of the molecular weight of peg molecules (8, 20 and 35 KDA) on cell function and allograft survival prolongation in pancreatic islets transplantation. Transplant Proc 38:2354–2355

80. Fang C, Shi B, Hong MH, Pei YY, Chen HZ (2006) Influence of particle size and MePEG molecular weight on in vitro macrophage uptake and in vivo long circulating of stealth nanoparticles in rats. Yao Xue Xue Bao 41:305–312

81. Dos Santos N, Allen C, Doppen AM, Anantha M, Cox KA, Gallagher RC, Karlsson G, Edwards K, Kenner G, Samuels L, Webb MS, Bally MB (2007) Influence of poly(ethylene glycol) grafting density and polymer length on liposomes: relating plasma circulation lifetimes to protein binding. Biochim Biophys Acta 1768:1367–1377

82. Zhang ZP, Lee SH, Feng SS (2007) Folate-decorated poly(lactide-co-glycolide)-vitamin E TPGS nanoparticles for targeted drug delivery. Biomaterials 28:1889–1899

83. Zhang ZP, Feng SS (2006) The drug encapsulation efficiency, in vitro drug release, cellular uptake and cytotoxicity of paclitaxel-loaded poly(lactide)-tocopheryl polyethylene glycol succinate nanoparticles. Biomaterials 27:4025–4033

84. Owens DE, Peppas NA (2006) Opsonization, biodistribution, and pharmacokinetics of polymeric nanoparticles. Int J Pharm 307:93–102

85. Bittner B, Morlock M, Koll H, Winter G, Kissel T (1998) Recombinant human erythropoietin (rhEPO) loaded poly(lactide-co-glycolide) microspheres: influence of the encapsulation technique and polymer purity on microsphere characteristics. Eur J Pharm Biopharm 45:295–305

86. Gaucher G, Dufresne MH, Sant VP, Kang N, Maysinger D, Leroux JC (2005) Block copolymer micelles: preparation, characterization and application in drug delivery. J Control Release 109:169–188

87. Allen C, Maysinger D, Eisenberg A (1999) Nano-engineering block copolymer aggregates for drug delivery. Colloide Surface B 16:3–27

88. Liu J, Zeng F, Allen C (2007) In vivo fate of unimers and micelles of a poly(ethylene glycol)-block-poly(caprolactone) copolymer in mice following intravenous administration. Eur J Pharm Biopharm 65:309–319

89. Zhang LF, Eisenberg A (1995) Multiple morphologies of crew-cut aggregates of Polystyrene-B-Poly(Acrylic Acid) block-copolymers. Science 268:1728–1731

90. Geze A, Putaux JL, Choisnard L, Jehan P, Wouessidjewe D (2004) Long-term shelf stability of amphiphilic beta-cyclodextrin nanosphere suspensions monitored by dynamic light scattering and cryo-transmission electron microscopy. J Microencapsul 21:607–613

91. Bittner B, Ronneberger B, Zange R, Volland C, Anderson JM, Kissel T (1998) Bovine serum albumin loaded poly(lactide-co-glycolide) microspheres: the influence of polymer purity on particle characteristics. J Microencapsul 15:495–514

92. Meinel L, Illi OE, Zapf J, Malfanti M, Peter Merkle H, Gander B (2001) Stabilizing insulin-like growth factor-I in poly(D,L-lactide-co-glycolide) microspheres. J Control Release 70:193–202

93. Perez C, De Jesus P, Griebenow K (2002) Preservation of lysozyme structure and function upon encapsulation and release from poly(lactic-co-glycolic) acid microspheres prepared by the water-in-oil-in-water method. Int J Pharm 248:193–206

94. Kang F, Jiang G, Hinderliter A, DeLuca PP, Singh J (2002) Lysozyme stability in primary emulsion for PLGA microsphere preparation: effect of recovery methods and stabilizing excipients. Pharm Res 19:629–633

95. Cao R, Villalonga R, Fragoso A (2005) Towards nanomedicine with a supramolecular approach: a review. IEE Proc Nanobiotechnol 152:159–164

96. Branchu S, Forbes RT, York P, Nyqvist H (1999) A central composite design to investigate the thermal stabilization of lysozyme. Pharm Res 16:702–708

97. Mohl S, Winter G (2006) Continuous release of Rh-interferon (alpha-2a from triglyceride implants: storage stability of the dosage forms. Pharm Dev Technol 11:103–110

98. Sanchez A, Tobio M, Gonzalez L, Fabra A, Alonso MJ (2003) Biodegradable micro- and nanoparticles as long-term delivery vehicles for interferon-alpha. Eur J Pharm Sci 18: 221–229

99. Van de Weert M, Hoechstetter J, Hennink WE, Crommelin DJ (2000) The effect of a water/organic solvent interface on the structural stability of lysozyme. J Control Release 68:351–359

100. Morlock M, Kissel T, Li YX, Koll H, Winter G (1998) Erythropoietin loaded microspheres prepared from biodegradable LPLG-PEO-LPLG triblock copolymers: protein stabilization and in-vitro release properties. J Control Release 56:105–115

101. Johansen P, Men Y, Audran R, Corradin G, Merkle HP, Gander B (1998) Improving stability and release kinetics of microencapsulated tetanus toxoid by co-encapsulation of additives. Pharm Res 15:1103–1110

102. Pean JM, Boury F, Venier-Julienne MC, Menei P, Proust JE, Benoit JP (1999) Why does PEG 400 co-encapsulation improve NGF stability and release from PLGA biodegradable microspheres? Pharm Res 16:1294–1299

103. Uchida T, Shiosaki K, Nakada Y, Fukada K, Eda Y, Tokiyoshi S, Nagareya N, Matsuyama K (1998) Microencapsulation of hepatitis B core antigen for vaccine preparation. Pharm Res 15:1708–1713

104. Wolf M, Wirth M, Pittner F, Gabor F (2003) Stabilisation and determination of the biological activity of L-asparaginase in poly(D,L-lactide-co-glycolide) nanospheres. Int J Pharm 256:141–152

105. Lam XM, Duenas ET, Cleland JL (2001) Encapsulation and stabilization of nerve growth factor into poly(lactic-co-glycolic) acid microspheres. J Pharm Sci 90:1356–1365

106. Perez-Rodriguez C, Montano N, Gonzalez K, Griebenow K (2003) Stabilization of alpha-chymotrypsin at the CH2Cl2/water interface and upon water-in-oil-in-water encapsulation in PLGA microspheres. J Control Release 89:71–85

107. Castellanos IJ, Crespo R, Griebenow K (2003) Poly(ethylene glycol) as stabilizer and emulsifying agent: a novel stabilization approach preventing aggregation and inactivation of proteins upon encapsulation in bioerodible polyester microspheres. J Control Release 88: 135–145

108. Van Eerdenbrugh B, Froyen L, Martens JA, Blaton N, Augustijns P, Brewster M, Van den Mooter G (2007) Characterization of physico-chemical properties and pharmaceutical performance of sucrose co-freeze-dried solid nanoparticulate powders of the anti-HIV agent loviride prepared by media milling. Int J Pharm 338:198–206

109. Michalet X, Pinaud FF, Bentolila LA, Tsay JM, Doose S, Li JJ, Sundaresan G, Wu AM, Gambhir SS, Weiss S (2005) Quantum dots for live cells, in vivo imaging, and diagnostics. Science 307:538–544

110. Hardman R (2006) A toxicologic review of quantum dots: toxicity depends on physicochemical and environmental factors. Environ Health Perspect 114:165–172

111. Derfus A, Chan W, Bhatia S (2003) Probing the cytotoxicity of semiconductor quantum dots. Nano Lett 4:11–18

112. Aldana J, Lavelle N, Wang Y, Peng X (2005) Size-dependent dissociation pH of thiolate ligands from cadmium chalcogenide nanocrystals. J Am Chem Soc 127:2496–2504

113. Aldana J, Wang YA, Peng X (2001) Photochemical instability of CdSe nanocrystals coated by hydrophilic thiols. J Am Chem Soc 123:8844–8850

114. Bianco A, Kostarelos K, Partidos CD, Prato M (2005) Biomedical applications of functionalised carbon nanotubes. Chem Comm 5:571–577

115. Bianco A, Kostarelos K, Prato M (2005) Applications of carbon nanotubes in drug delivery. Curr Opin Chem Biol 9:674–679

116. Sayes CM, Liang F, Hudson JL, Mendez J, Guo W, Beach JM, Moore VC, Doyle CD, West JL, Billups WE, Ausman KD, Colvin VL (2006) Functionalization density dependence of single-walled carbon nanotubes cytotoxicity in vitro. Toxicol Lett 161:135–142

117. Donaldson K, Aitken R, Tran L, Stone V, Duffin R, Forrest G, Alexander A (2006) Carbon nanotubes: a review of their properties in relation to pulmonary toxicology and workplace safety. Toxicol Sci 92:5–22

118. Donaldson K, Stone V, Tran CL, Kreyling W, Borm PJ (2004) Nanotoxicology. Occup Environ Med 61:727–728

119. Dufes C, Uchegbu IF, Schatzlein AG (2005) Dendrimers in gene delivery. Adv Drug Deliv Rev 57:2177–2202

120. Borm PJ (2002) Particle toxicology: from coal mining to nanotechnology. Inhal Toxicol 14:311–324

121. Roberts JC, Bhalgat MK, Zera RT (1996) Preliminary biological evaluation of polyamidoamine (PAMAM) Starburst dendrimers. J Biomed Mater Res 30:53–65

122. Malik N, Wiwattanapatapee R, Klopsch R, Lorenz K, Frey H, Weener JW, Meijer EW, Paulus W, Duncan R (2000) Dendrimers: relationship between structure and biocompatibility in vitro, and preliminary studies on the biodistribution of 125I-labelled polyamidoamine dendrimers in vivo. J Control Release 65:133–148

123. Hong S, Bielinska AU, Mecke A, Keszler B, Beals JL, Shi X, Balogh L, Orr BG, Baker JR, Jr., Banaszak Holl MM (2004) Interaction of poly(amidoamine) dendrimers with supported lipid bilayers and cells: hole formation and the relation to transport. Bioconjug Chem 15:774–782

124. Chen HT, Neerman MF, Parrish AR, Simanek EE (2004) Cytotoxicity, hemolysis, and acute in vivo toxicity of dendrimers based on melamine, candidate vehicles for drug delivery. J Am Chem Soc 126:10044–10048

125. Lee JH, Lim YB, Choi JS, Lee Y, Kim TI, Kim HJ, Yoon JK, Kim K, Park JS (2003) Polyplexes assembled with internally quaternized PAMAM-OH dendrimer and plasmid DNA have a neutral surface and gene delivery potency. Bioconjug Chem 14:1214–1221

126. Schatzlein AG, Zinselmeyer BH, Elouzi A, Dufes C, Chim YT, Roberts CJ, Davies MC, Munro A, Gray AI, Uchegbu IF (2005) Preferential liver gene expression with polypropylenimine dendrimers. J Control Release 101:247–258

127. Choi JS, Nam K, Park JY, Kim JB, Lee JK, Park JS (2004) Enhanced transfection efficiency of PAMAM dendrimer by surface modification with L-arginine. J Control Release 99:445–456

128. Godbey WT, Wu KK, Hirasaki GJ, Mikos AG (1999) Improved packing of poly(ethylenimine)/DNA complexes increases transfection efficiency. Gene Ther 6:1380–1388

129. Roessler BJ, Bielinska AU, Janczak K, Lee I, Baker JR, Jr. (2001) Substituted beta-cyclodextrins interact with PAMAM dendrimer-DNA complexes and modify transfection efficiency. Biochem Biophys Res Commun 283:124–129

130. Blanco D, Alonso MJ (1998) Protein encapsulation and release from poly(lactide-co-glycolide) microspheres: effect of the protein and polymer properties and of the co-encapsulation of surfactants. Eur J Pharm Biopharm 45:285–294

131. Kwon YM, Baudys M, Knutson K, Kim SW (2001) In situ study of insulin aggregation induced by water-organic solvent interface. Pharm Res 18:1754–1759

132. Raghuvanshi RS, Goyal S, Singh O, Panda AK (1998) Stabilization of dichloromethane-induced protein denaturation during microencapsulation. Pharm Dev Technol 3:269–276

133. Griebenow K, Klibanov AM (1996) On protein denaturation in aqueous-organic mixtures but not in pure organic solvents. J Am Chem Soc 118:11695–11700

134. Zambaux MF, Bonneaux F, Gref R, Dellacherie E, Vigneron C (1999) Preparation and characterization of protein C-loaded PLA nanoparticles. J Control Release 60:179–188

135. Shoyele SA, Cawthorne S (2006) Particle engineering techniques for inhaled biopharmaceuticals. Adv Drug Deliv Rev 58:1009–1029

136. Johnson KA (1997) Preparation of peptide and protein powders for inhalation. Adv Drug Deliv Rev 26:3–15

137. Schlocker W, Gschliesser S, Bernkop-Schnurch A (2006) Evaluation of the potential of air jet milling of solid protein-poly(acrylate) complexes for microparticle preparation. Eur J Pharm Biopharm 62:260–266

138. Hinrichs WL, Sanders NN, De Smedt SC, Demeester J, Frijlink HW (2005) Inulin is a promising cryo- and lyoprotectant for PEGylated lipoplexes. J Control Release 103:465–479

139. Ohtake S, Schebor C, Palecek SP, de Pablo JJ (2005) Phase behavior of freeze-dried phospholipid-cholesterol mixtures stabilized with trehalose. Biochim Biophys Acta 1713:57–64

140. Bittner B, Kissel T (1999) Ultrasonic atomization for spray drying: a versatile technique for the preparation of protein loaded biodegradable microspheres. J Microencapsul 16:325–341

141. Adler M, Unger M, Lee G (2000) Surface composition of spray-dried particles of bovine serum albumin/trehalose/surfactant. Pharm Res 17:863–870

142. Codrons V, Vanderbist F, Verbeeck RK, Arras M, Lison D, Preat V, Vanbever R (2003) Systemic delivery of parathyroid hormone (1-34) using inhalation dry powders in rats. J Pharm Sci 92:938–950

143. Andya JD, Maa YF, Costantino HR, Nguyen PA, Dasovich N, Sweeney TD, Hsu CC, Shire SJ (1999) The effect of formulation excipients on protein stability and aerosol performance of spray-dried powders of a recombinant humanized anti-IgE monoclonal antibody. Pharm Res 16:350–358

144. Bosquillon C, Preat V, Vanbever R (2004) Pulmonary delivery of growth hormone using dry powders and visualization of its local fate in rats. J Control Release 96:233–244

145. Bosquillon C, Rouxhet PG, Ahimou F, Simon D, Culot C, Preat V, Vanbever R (2004) Aerosolization properties, surface composition and physical state of spray-dried protein powders. J Control Release 99:357–367

146. Labrude P, Rasolomanana M, Vigneron C, Thirion C, Chaillot B (1989) Protective effect of sucrose on spray drying of oxyhemoglobin. J Pharm Sci 78:223–229

147. Liao YH, Brown MB, Jones SA, Nazir T, Martin GP (2005) The effects of polyvinyl alcohol on the in vitro stability and delivery of spray-dried protein particles from surfactant-free HFA 134a-based pressurised metered dose inhalers. Int J Pharm 304:29–39

148. Brandenberger H (2003) Best@buchi evaporation. Inf Bull 27

149. Irngartinger M, Camuglia V, Damm M, Goede J, Frijlink HW (2004) Pulmonary delivery of therapeutic peptides via dry powder inhalation: effects of micronisation and manufacturing. Eur J Pharm Biopharm 58:7–14

150. Chan HK (2006) Dry powder aerosol delivery systems: current and future research directions. J Aerosol Med 19:21–27
151. Costantino HR, Firouzabadian L, Hogeland K, Wu C, Beganski C, Carrasquillo KG, Cordova M, Griebenow K, Zale SE, Tracy MA (2000) Protein spray-freeze drying. Effect of atomization conditions on particle size and stability. Pharm Res 17:1374–1383
152. Yu Z, Johnston KP, Williams RO, 3rd (2006) Spray freezing into liquid versus spray-freeze drying: influence of atomization on protein aggregation and biological activity. Eur J Pharm Sci 27:9–18
153. Yu Z, Garcia AS, Johnston KP, Williams RO, 3rd (2004) Spray freezing into liquid nitrogen for highly stable protein nanostructured microparticles. Eur J Pharm Biopharm 58:529–537
154. Yu Z, Rogers TL, Hu J, Johnston KP, Williams RO, 3rd (2002) Preparation and characterization of microparticles containing peptide produced by a novel process: spray freezing into liquid. Eur J Pharm Biopharm 54:221–228
155. Rogers TL, Hu J, Yu Z, Johnston KP, Williams RO, 3rd (2002) A novel particle engineering technology: spray-freezing into liquid. Int J Pharm 242:93–100
156. Rogers TL, Nelsen AC, Hu J, Brown JN, Sarkari M, Young TJ, Johnston KP, Williams RO, 3rd (2002) A novel particle engineering technology to enhance dissolution of poorly water soluble drugs: spray-freezing into liquid. Eur J Pharm Biopharm 54:271–280
157. Henczka M, Baldyga J, Shekunov BY (2006) Modelling of spray-freezing with compressed carbon dioxide. Chem Eng Sci 61:2880–2887
158. Pasquali I, Bettini R, Giordano F (2006) Solid-state chemistry and particle engineering with supercritical fluids in pharmaceutics. Eur J Pharm Sci 27:299–310
159. Elvassore N, Baggio M, Pallado P, Bertucco A (2001) Production of different morphologies of biocompatible polymeric materials by supercritical $CO(2)$ antisolvent techniques. Biotechnol Bioeng 73:449–457
160. NIOSH (2006) Approaches to Safe nanotechnology: An Information Exchange with NIOSH. The National Institute of Occupational Safety and Health, Washington, DC
161. Oberdorster G, Maynard A, Donaldson K, Castranova V, Fitzpatrick J, Ausman K, Carter J, Karn B, Kreyling W, Lai D, Olin S, Monteiro-Riviere N, Warheit D, Yang H (2005) Principles for characterizing the potential human health effects from exposure to nanomaterials: elements of a screening strategy. Part Fibre Toxicol 2:8
162. Oberdorster G, Oberdorster E, Oberdorster J (2005) Nanotoxicology: an emerging discipline evolving from studies of ultrafine particles. Environ Health Perspect 113:823–839
163. Curtis J, Greenberg M, Kester J, Phillips S, Krieger G (2006) Nanotechnology and nanotoxicology: a primer for clinicians. Toxicol Rev 25:245–260
164. Maynard AD, Kuempel ED (2005) Airborne nanostructured particles and occupational health. J Nanopart Res 7:587–614
165. Nel A, Xia T, Madler L, Li N (2006) Toxic potential of materials at the nanolevel. Science 311:622–627
166. Oberdorster G, Ferin J, Gelein R, Soderholm SC, Finkelstein J (1992) Role of the alveolar macrophage in lung injury: studies with ultrafine particles. Environ Health Perspect 97: 193–199
167. Li N, Sioutas C, Cho A, Schmitz D, Misra C, Sempf J, Wang M, Oberley T, Froines J, Nel A (2003) Ultrafine particulate pollutants induce oxidative stress and mitochondrial damage. Environ Health Perspect 111:455–460
168. MacNee W, Donaldson K (2003) Mechanism of lung injury caused by PM10 and ultrafine particles with special reference to COPD. Eur Respir J 40:47s–51s
169. Shvedova AA, Castranova V, Kisin ER, Schwegler-Berry D, Murray AR, Gandelsman VZ, Maynard A, Baron P (2003) Exposure to carbon nanotube material: assessment of nanotube cytotoxicity using human keratinocyte cells. J Toxicol Environ Health A 66:1909–1926
170. Monteiro-Riviere NA, Nemanich RJ, Inman AO, Wang YY, Riviere JE (2005) Multi-walled carbon nanotube interactions with human epidermal keratinocytes. Toxicol Lett 155: 377–384

171. Anand VP. Biocompatibility Safety Assessment of Medical Devices: FDA/ISO and Japanese Guidelines. http://www.devicelink.com/mddi/archive/00/01/017.html
172. FDA (2004) Challenge and opportunity on the critical path to new medical products. US Food and Drug Administration, Rockville, MD
173. FDA (2006) Early Development considerations for innovative combination products. US Food and Drug Administration, Rockville, MD
174. FDA (2006) Exploratory IND studies. US Food and Drug Administration, Rockville, MD
175. FDA (2001) Changes or modifications during the conduct of a clinical investigation; Final guidance for industry and CDRH staff. US Food and Drug Administration, Rockville, MD

Chapter 5
Biomedical Applications of Nanoparticles

G.L. Prasad

Abstract Nanomaterials hold immense promise for significantly improving existing diagnosis, therapy and designing novel approaches to treat a variety of human
ailments. While some of the applications of nanotechnology have been translated
into clinical settings, many more potential uses of nanomedicines have been demonstrated in experimental systems. Since a variety of materials can be nanosized, the
scope of nanomedicine is also large and may even become larger. At the same time,
the impact of nanomaterials on cellular and animal models will need to be carefully
evaluated under both acute and chronic exposures at toxicological and pharmacological doses. It is extremely important to evaluate the basic issues such as the fate
of nanomaterials in biological systems, and how the cells and tissues react to the
exposure of nanomaterials in developing nanomedicines. This chapter will cover
some of these promises and future directions.

Contents

G.L. Prasad (✉)
Department of Physiology, Temple University School of Medicine, 3400 North Broad Street,
OMS228, Philadelphia, PA 19140
e-mail: glprasad@temple.edu

T.J. Webster (ed.), *Safety of Nanoparticles*, Nanostructure Science and Technology,
DOI 10.1007/978-0-387-78608-7_5, © Springer Science+Business Media, LLC 2009

5.1 Introduction

Nanomaterials, due to their unique physical, mechanical, and chemical properties, are used in electronics and numerous consumer products. A diverse array of engineered nanomaterials is being produced in hundreds of tons a year. The commercial value of the nanotechnology industry, by some estimates, is anticipated to exceed several hundreds of billions of dollars annually. Over the past few years, it has become increasingly clear that nano-sized materials offer immense potential for various biomedical applications (Table 5.1). Consistent with this enthusiasm, a large

Table 5.1 Potential biomedical applications of some nanoparticles

Diagnosis and sensory applications		
Material	Application	Comments
Quantum dots	Imaging and tracking by fluorescence	Exhibit broad emission spectra; can be coupled to antibodies; toxicity a concern
Iron oxide particles	Imaging and tracking by MRI	Useful as multifunctional particles; can be used to generate local heating and kill targeted tissue
Nanotubes	Biomolecular sensors	Cell tracking, optical labeling, MRI contrast agent, Electrocatalysis (e.g., glucose sensing)
Gold shells	Biomolecular sensors	Electrocatalysis (e.g., glucose sensing)
Therapeutic applications		
C_{60} fullerenes	Drug carriers	Anti-oxidants and singlet oxygen generator can be coupled to targeting molecules and antimicrobial agents
Nanotubes	Drug and gene delivery, imaging and photothermal ablation of cells	Highly versatile particles can be used as multifunctional agents for a variety of applications
Liposomes	Drug delivery	Versatile biodegradable multifunctional particles, clinically approved
Dendrimers and polymers	Drug delivery	Biodegradable polymers, slow release of drugs, alternatives for injectable drugs
Gold nanoshells	Drug delivery	Photothermal ablation of targeted tissue can be used as reporters

Table 5.2 Web resources: a selected list of web resources that cover biomedical aspects nanotechnology. Resources such as the nanotechweb and Foresight Institute send a weekly electronic newsletter for registrants

Selected web resources and news letters	
Nanotechnology Initiative (NIH)	http://nano.gov/
Institute of Physics (UK)	http://nanotechweb.org
Foresight Nanotech Institute	http://www.foresight.org/
Nanoscale Informal Science Education	http://qt.exploratorium.edu/
Network (NISE)	nise-resources/index.php
NanoEd Resource Portal	http://www.nanoed.org/

number of publications and reviews covering therapeutic, imaging, reporting, and sensory applications of C_{60} fullerenes, nanotubes, quantum dots, and other novel-engineered materials have appeared. Several reviews [1, 2, 3, 4, 5, 6, 7, 8, 9, 10], newsletters, and websites (Table 5.2) also describe and highlight advances in nanotechnology on a regular basis. In addition, other types of nano-sized materials such as drug formulations including liposomal, dendrimeric, and other nanosized drugs have also been pursued and determined clinically useful. The purpose of this chapter is to cover the general aspects of biomedical applications, including some limitations, of nanotechnology. Therefore, readers should refer to specific and original research papers or reviews for an in depth discussion of a given application of a nanoparticle.

Nanomaterials used for biomedical applications are composed of chemically diverse compounds typically less than 100 nm in size. However, some investigators also refer to polymeric assemblies in the submicron (<1000 nm) size range as nanomaterials. Some of the key advantages of nanomedicine over the traditional approaches include enhancing the bioavailability of nanosized drugs and the opportunity to target and create multifunctional therapeutic agents. While the development of "nano-machines" and "nano-robots" that are capable of performing multiple tasks has been envisaged, the design and testing of such machines requires further research [11, 12]. At present, there are numerous examples of the effective use of nano-sized materials and some of them are in advanced clinical trials.

5.2 Nanomaterials of Biomedical Interest

Nanomaterials are comprised of chemically diverse and heterogeneous compounds. They may be engineered from pure carbon, inorganic or polymeric compounds. Although a wide spectrum of applications of nanomaterials has been proposed, significant clinical applications have been accomplished with polymeric nanomaterials that include biodegradable polymeric materials and liposomal preparations. Biomedical applications of engineered carbon particles and inorganic materials are being actively pursued, and require further validation before they can be used in a pre-clinical setting. Thus, nanomedicine, as a field, may be considered as an emerging area.

The properties and reactivity of nanomaterials significantly differ from their macromolecular counterparts. This distinctive property of nanoparticles is a consequence of the display of more atoms on their surfaces as the particle size decreases which is due to increased surface area. The molecules displayed on the surfaces exponentially increase as the particle size decreases to <100 nm [13]. Consequently, nanomaterials are likely to be more reactive, exhibit higher surface charge and agglomerate into larger clusters or react with other materials. The size of the nanoparticles is also an important determinant of their physicochemical properties such as solubility, optical, and catalytic behavior. The surface reactivity of nanoparticles is often modified by coating with various materials for stability and biocompatibility, and hence surface modifications are among the critical determinants of the usage of nanomaterials. Whereas 'pristine' (unmodified) nanomaterials (such as carbon nanotubes and fullerenes) have been synthesized and their properties have been investigated in depth, advances in chemistry of nanomaterials have enabled researchers to attach a large array of diverse molecules including proteins, peptides, DNA, drugs, and carbohydrates. The ability to conjugate such functionalities has heralded the area of nanomedicine, and is likely to result in creating multifunctional nanoparticles. Here, we will consider some of the more established nanomaterials and their respective biomedical applications.

5.2.1 Fullerenes

C_{60} fullerenes, popularly known as Bucky balls [1, 14] are icosahedral (20 sided) structures made of 60 carbon atoms (M_r 720), organized in a spherical arrangement of hexagonal and pentagonal groups. C_{60} molecules are 0.7 nm in size and can aggregate into larger crystalline structures, depending upon the medium in which they are suspended [15].

Some of the interesting properties of C_{60} include large electronegativity (high electron affinity). Second, crystals of fullerenes have strong photoluminescence (PL), which is sensitive to the exact "inter-ball" spacing, and this spacing is sensitive to the local pH and oxygen content. Third, fullerenes are amenable for chemical modifications which allow for the attachment of organic molecules of interest. Various types of C_{60} functionalization that enhance the hydrophilicity of fullerenes have been described (reviewed in [1]). Hence, multiple reactive groups and functionalities, including antibodies, drugs, and metals may be attached to the core of C_{60} particles. Fourth, some studies indicate that fullerenes may selectively accumulate in tumors and are reported to be excreted in urine without accumulating in the tissues [16]. Some studies also have reported that fullerenes cross and protect the blood brain barrier [17, 18], a property that can be exploited for certain applications.

Among the various properties of fullerenes, redox properties (electron acceptor–donor) have been extensively studied and are particularly attractive for biomedical

Fig. 5.1 Redox properties of C_{60} fullerenes

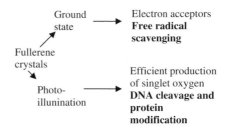

applications. Theoretically, C_{60} could accept up to 6 electrons. Due to their high affinity for electrons, fullerenes, in their ground state, are highly efficient free radical scavengers and are often referred to as free radical sponges (Fig. 5.1) [1, 14]. Several groups of researchers have demonstrated that C_{60} functions as a free radical scavenger in cultured cells and animal models. For example, neuroprotective effects [19, 20] and anti-apoptotic protective effects on UVB-irradiated keratinocytes [21] of carboxyfullerenes are believed to be due to the ability of C_{60} to scavenge free radicals. Further, derivatives of C_{60} are reported to protect against oxidative stress in cultured cells as well as rat ischemia-reperfused lungs [20], and that induced by carbon tetrachloride [22]. Further, several fullerene derivatives have been shown to effectively suppress the formation of hydroxyl, superoxide and ter-Butryl hydroperoxide-generated reactive oxygen species [23].

Fullerenes are efficiently converted to triplet states by UV and visible irradiation, and the triplet states efficiently sensitize the generation of singlet oxygen which is a highly reactive molecule [24]. Although there have been suggestions that this property may be used for a variety of therapies [1], a significant amount of work and rigorous evaluation of this evidence is necessary in this regard.

Advances in fullerene chemistry have allowed the introduction of numerous types of functionalities. For example, chemotherapeutic drugs such as taxol have been attached to the fullerene cage [25]. The advantages of using fullerenes as drug carriers may include improved bioavailability and tissue uptake of drugs compared to the free drugs. Fullerenes also have been modified to incorporate imaging agents such as gadolinium [26]. Since fullerenes have been found to "fit" into the active site of retroviral reverse transcriptase, they have been suggested as potential inhibitors of HIV [1]. While these and other beneficial properties of fullerene derivatives have been investigated in limited cell culture studies, more extensive evaluation in appropriate cellular and animal models is necessary.

5.2.2 Carbon Nanotubes

Nanotubes are another type of carbon nanoparticles that have been considered for a variety of biological applications, including as biosensors, drug delivery devices, and as therapeutic agents [27]. *Carbon nanotubes* are primarily composed of carbon atoms, arranged in benzene rings which form graphene sheets, and are rolled into tubes [28]. The single-walled carbon nanotubes (SWNT) consist of a single layer of graphene sheet, which are characterized by a high ratio of length over diameter (known as the aspect ratio). Thus, SWNTs can be several microns long with a diameter in the range of a few nanometers. Multi-walled nanotubes (MWNTs), on the other hand, consist of several concentric sheets, spaced at less than 1 nm. Nanotubes, because of their distinctive electrical, mechanical, and optical properties, are well suited for several biological applications. For example, nanotubes exhibit outstanding structural flexibility and fluidity and yet possess high mechanical strength. Nanotubes exhibit useful Raman scattering and fluorescence emission in the near infrared (nIR) spectrum between 900 and 1300 nm.

However, nanotubes in their unmodified state, like other carbon nanomaterials, are poorly soluble in aqueous media and hence are incompatible with biological media. Covalent and non-covalent modifications of nanotubes improve their compatibility with biological media. For example, biopolymers such as DNA (both single- and double-stranded DNA) and amphiphilic peptides wrap around the nanotubes and the complexes exhibit sufficient solubility in aqueous media [6, 29]. Both SWNTs and MWNTs can be efficiently functionalized nanotubes either through oxidative treatment using strong acids or through 1, 3-dipolar cycloaddition. Such modifications have allowed further introduction of a wide range of functionalities, including proteins, nucleotides, and other small molecular weight compounds. Additionally, open-ended cylindrical carbon nanotubes have been shown to accommodate smaller molecules within their internal cavities. Further, the sidewalls and the ends can be differentially functionalized as they are known to exhibit different chemical reactivities, thus providing greater flexibility in the design of nanotube-based drug delivery vehicles.

Owing to their electronic structure, optical properties in the nIR region and large surface area for adsorption of chemicals or biological compounds, nanotubes are considered to be useful electrochemical sensors for sensing key intracellular parameters in real time [6, 30]. Since, nanotubes fluoresce in the nIR region, the signal can penetrate tissues without interference from the autofluorescence from tissues. In the case of SWNTs, the sidewall modification is accomplished through non-covalent functionalization which is necessary to preserve optical properties. The potential of nanotubes as biological sensors has been amply demonstrated for monitoring β-D-glucose concentrations that are physiologically relevant [31]. SWNTs are functionalized with glucose oxidase, and ferricyanide is irreversibly adsorbed onto the nanotubes which acts as an electron-withdrawing group and thus quenches the inherent fluorescence signal of the nanotubes. The interaction of glucose oxidase with its substrate, glucose, produces hydrogen peroxide, which in turn, reduces ferricyanide resulting in the transfer of electrons and increases the fluorescent signal of nanotubes. It has also been shown that the detection of the activity of glucose oxidase and other related biocatalytic systems has been accomplished by amperometric techniques with higher sensitivity. The high electrical conductivity of nanotubes, at least in part, is hypothesized to account for the enhanced sensitivity of nanotube biosensors [27].

The superior mechanical strength of nanotubes may be exploited for various applications in tissue engineering [6]. Whereas commonly used synthetic polymers such as PLGA and PLA are biodegradable, they lack desirable mechanical strength. In this context, the necessary structural reinforcement may be provided by carbon nanotubes. However, it should be pointed out that nanotubes are not biodegradable and hence are likely to persist in tissues. By employing substantially smaller quantities of nanotubes in the nanocomposites, it may be possible to reduce any potentially hazardous effects of the nanomaterials and yet achieve the desired beneficial effects of nanotubes. Several studies have shown that nanotubes support growth of smooth muscle cells, fibroblasts, and neurons. Additionally, nanotube-containing composites have been reported to stimulate bone formation.

Functionalized nanotubes have been shown to reach intracellular compartments, most likely through endocytosis [32]. These findings suggested that nanotubes might serve as novel drug delivery vehicles. Indeed, some studies demonstrate that nanotubes functionalized with antifungal drugs such as amphotericin B efficiently transport the drug to mammalian cells. Further, nanotubes have been employed as gene transfection vectors [29]. While some studies show that transfection of cells is achieved through nanotubes, their utility for gene therapy in animal models remains to be tested. Additionally, nanotubes may be used for photodynamic/photothermal irradiation therapy of targeted tumor cells. In this strategy, nanotubes are targeted to tumor tissue through a couple targeting molecules (such as tumor-specific antibody or a small molecular compound) and are irradiated with an nIR laser. The nanotubes absorb nIR energy, transduce the energy and cause local heating which kills the targeted tumor tissues [32]. Whereas this approach is attractive, several issues including the amount of irradiation, the number of nanotubes necessary for cell killing, and the biodistribution of the nanotubes require additional consideration.

5.2.3 Quantum Dots

Inorganic nanoparticles such as semiconductor nanocrystals, commonly known as quantum dots, have found numerous applications in biology and medicine as potential imaging agents [23, 33]. Typically inorganic nanoparticles contain a central core that defines the reporting properties of the particle, which is surrounded by an organic coating that protects the core from physiological environments; further, several functionalities such as proteins and oligonucleotides have been linked to the organic layer for targeting the nanoparticle (Fig. 5.2). Depending on the nature of the central core, inorganic nanoparticles may serve as fluorescent probes (quantum dots), Raman probes or as contrast enhancement agents to improve the sensitivity of magnetic resonance imaging (MRI).

Quantum dots are fluorescent nanoparticles of about 2–10 nm in size [33]. The central core is composed of hundreds to thousands of atoms belonging to the groups of II–VI (e.g., cadmium, selenium) or III–V (e.g., indium) elements. The fluorescence emission spectrum of quantum dots is dependent on the size of the central

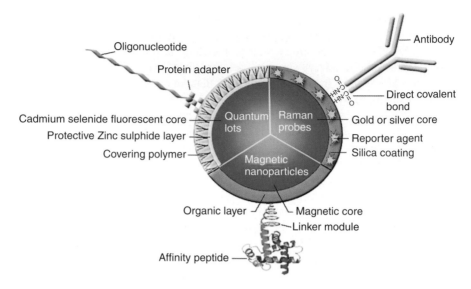

Fig. 5.2 A schematic representation of inorganic nanoparticles (reprinted from [23])

nanocrystal core which can be adjusted by controlling the amount of precursors during the synthesis. The modest quantum yield of the core nanocrystal is substantially (up to 80%) enhanced by wrapping the core in a zinc sulfide shell. Since the core-shell of the nanocrystal is hydrophobic, it is rendered water soluble with bifunctional molecules (e.g., mercaptohydrocarbonic acid), surface silanization or coating the surface with amphiphilic polymers. The quantum dot crystals are functionalized by linking appropriate targeting molecules through either covalent or non-covalent methods.

Quantum dots offer significant advantages over the conventional organic fluorochromes. For example, quantum dots can be tuned to emit brighter fluorescence from ultraviolet to nIR ranges (450–850 nm) with a narrow spectral emission peaks and minimum overlap between the spectra. Compared to the organic dyes, quantum dots have broader absorption spectra which enable them to be excited over a broad range of wavelengths. These properties make quantum dots ideal probes for detection of multicolor imaging. Their extreme brightness coupled with the resistance to photobleaching allows for their examination at relatively low laser intensities over a long period, which is particularly suited for live-cell imaging. Notwithstanding these useful properties, quantum dots also possess some disadvantages. Quantum dots are known to "blink" – a property of quantum dots that describes alternating between emitting and non-emitting states.

The strong fluorescence, photostability, tunability, and narrow emission spectra of quantum dots led to the development of numerous biological applications, wherein the nanoparticles are used as probes [34]. Most commonly, quantum dots are conjugated to proteins, nucleic acids, or other biologically relevant molecules which serve as molecular targeting devices. Therefore, quantum dots can be used in diagnosis and molecular detection, profiling, signal transduction studies, and biomedical imaging. A number of biologically important molecular targets, including tumor markers (HER2/neu oncogene) have been detected in immunohistochemistry and immunoblotting methods. Several groups of researchers have taken advantage of the size-dependent emission spectra of quantum dots and detected multiple molecular targets [23, 35]. Thus, quantum dot conjugates should be applicable for routine pathological evaluation of clinical samples. Further, quantum dots of defined sizes (5–30 nm) have been suggested to be useful tools for mapping the lymphatic system and sentinel lymph nodes, which are important in cancer therapy.

5.2.4 Superparamagnetic Nanoparticles

Superparamagnetic nanoparticles contain a magnetically active metal core and are widely used in MRI imaging for enhancing contrast [3]. Iron oxide containing superparamagnetic nanoparticles has been commonly used as the core particle. The size and charge of iron oxide crystals and the nature of chemical coatings are some of the key determinants of the utility of these nanoparticles. Because of the excellent spatial resolution afforded by MRI, iron oxide nanoparticles have been used for diagnostic imaging of tumors, inflammatory and degenerative diseases. Iron oxide particles also have been conjugated to numerous antibodies for targeted molecular and cellular imaging. In addition, iron oxide nanoparticles have been used to label cells and track marked cells to determine the in vivo fate of the transplanted cells for future cellular therapies, and appropriate protocols for generating magneto-fluorescent imaging particles have been recently described [36]. A more novel approach is to employ iron oxide particles for magnetic induction of hyperthermia. Tissues containing targeted magnetic particles are subjected to a magnetic field, "heating up" the tissues to 43–45°C which results in selective cell killing [37]. *Raman probes*, which contain gold or silver as the metallic core, a reporter molecule with a spectroscopic signature and silica shell for protein conjugation, are also potentially useful ultra sensitive biological probes [23].

5.2.5 Polymeric Micelles and Liposomes

Among various nanoparticles, *polymeric micelles* and *liposomes* have been extensively investigated and are particularly well suited for drug delivery purposes. Polymeric micelles contain a hydrophobic core for accommodating hydrophobic drugs, which is surrounded by hydrophilic polymers (Fig. 5.3 [9]). Polymeric micelles, which range in 20–250 nm in size, are typically assembled through biocompatible block amphiphilic copolymers. Among the advantages of employing polymeric micelles for drug delivery are relatively simple preparation, efficient drug loading

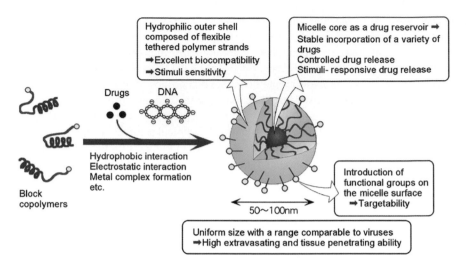

Fig. 5.3 Design of polymeric micelles as drug and gene delivery vehicles (reprinted from [9])

with minimal modification and controlled drug delivery. Several drug formulations of commonly used chemotherapeutic drugs such as doxorubicin and Paclitaxel are currently in various phases of testing and clinical trials.

Liposomes, on the other hand, are spherically closed structures formed by phospholipid bilayers with an aqueous phase inside and between lipid bilayers [38]. The sizes of liposomes are variable with unilamellar liposomes smaller (100–800 nm) than multilamellar liposomes (up to 5000 nm), which consist of several layers of concentric lipid bilayers. Both hydrophilic and hydrophobic drugs can be trapped in liposomes, and the enclosed drugs are generally protected from the external environment with longer circulation times (enhanced bioavailability). By varying the size and composition of liposomes it is possible to control the properties of liposomes. Further, the basic liposomes can be modified by introducing antibodies or small molecules such as folate for targeting tumors or may be used for diagnostic imaging. Currently, several important drugs modified as liposomal formulations are undergoing clinical evaluation. For example, polyethylene glycol-doxorubicin liposomes, known as Doxil, are currently used to treat several cancers. Treatment with Doxil has been shown to reduce cardiotoxicity generally associated with doxorubicin treatment.

5.3 Factors that Require Considerations for Using Nanomaterials for Biomedical Applications

While it is abundantly clear that nanomaterials hold immense promise in the diagnosis and treatment of a number of human ailments, several concerns need to be addressed first. They include potential toxicity, metabolism, stability, clearance mechanisms, and in vivo detection. Additionally, one must take into account the parameters that may enhance or limit tissue uptake of nanomaterials.

5.3.1 Nanotoxicity

Exposure of naturally occurring airborne nanoparticles to humans has been known to occur for a long time. However, exposure to nanomaterials from industrial activity coupled with a profound increase in the synthesis of a wide array of nanomaterials and their uses raises particular safety concerns [13, 39, 40]. The unique properties of nanomaterials (surface area and size distribution, chemical properties, and surface structure) could elicit different, and yet potent, biological responses from their macro-sized counterparts [12]. Indeed, these are properties which make nanomaterials unique and attractive for biomedical applications. Some properties of nanomaterials that could influence their toxicity are listed in Table 5.3. Considering the ever increasing use of nanomaterials in various industrial and consumer products, and the promise they hold in improving diagnosis and treatment of various human ailments, it is necessary to evaluate their potentially hazardous effects. The exposure of nanoparticles to humans may conceivably come through the following paths: accidental exposure, environmental exposure arising from industrial waste discharge, exposure from consumer goods consisting of nanomaterials, or through administration of nanomedicines.

Nanomaterials may gain entry into human systems through inhalation, ingestion, or through dermal routes. Deposition and the physiological consequences elicited by inhaled nanoparticles are dependent on the size, shape (e.g., spherical, fibrous), surface charge, and the aggregation state of the nanoparticles. For example, smaller (<100 nm) particles are likely to travel deeper into the lung and deposit into the alveolar region. Particles deposited in the upper airways are cleared mainly through the mucociliary escalator [13, 39], whereas materials deposited in the deeper regions including alveoli may translocate to extrapulmonary sites through a variety of mechanisms across the pulmonary epithelium, entering systemic circulation. Such translocated nanomaterials may exert deleterious effects on the cardiovascular system. The observed adverse effects are hypothesized to result from the inflammatory responses elicited by the persistence of pulmonary nanoparticles or the systemic effects (hemostasis or cardiovascular integrity). Further, inhaled manganese

Table 5.3 General characteristics of nanoparticles which determine their toxicity (summarized from [40])	Particle number and size distribution Dose of particles to target tissue Surface chemistry Aggregation/agglomeration of engineered nanoparticles Surface charge Particle shape and/or electrostatic attraction potential Method of particle synthesis and post-synthetic modification

dioxide nanoparticles or intranasally deposited gold nanoparticles have been shown to translocate to olfactory bulbs and localize to mitochondria.

In contrast to the studies on inhaled nanoparticles, there is limited information on how ingested materials or those entering the gastro intestinal system, and it appears to depend upon the composition and size of the particles. Whereas most ingested nanoparticles are excreted in feces, a notable quantity of smaller nanoparticles has been reported to re-enter systemic circulation and organs. For example, whereas a large portion of orally administered C_{60} fullerenes to rats are excreted through feces, some of the ingested material is re-absorbed. Similarly, nanosized titanium oxide and polysterene beads are found to be taken up from the intestine in a size-dependent fashion. However, radioactive colloids (technetium-99m labeled sulfur colloid) have been orally administered and imaged through the GI tract. These colloidal particles (10–900 nm) have not been known to absorb from the GI tract, but are excreted in the feces (Personal communication, Dr. L. C. Knight, Temple University School of Medicine). Therefore, size alone may not determine the uptake or excretion of nanosized materials.

Absorption through the skin is another important route of exposure of nanomaterials. Considering that titanium and zinc oxide nanoparticles are used as components of sunscreens and lotions, several studies were performed to test whether they penetrate skin. In the case of titanium oxide nanoparticles, the method of administration (emulsion) appears to influence dermal absorption. Further, evidence was presented that hair follicles may in fact serve as a repository of topically applied nanoparticles. The available literature also suggests that uptake of nanoparticles through the skin is size dependent, with nano-sized particles more likely to penetrate deeper than the larger particles.

5.3.2 Cellular Responses to Nanoparticle Exposure

Given that nanoparticles markedly differ in their physicochemical properties, they are anticipated to elicit a wide range of cellular responses. Since cellular responses to nanoparticles (or any other agent) are measured under controlled conditions, among other numerous factors, particular attention must be paid to the quality of the nanomaterials preparation, use of appropriate cell type, culture conditions, and interpretation of the experimental results. It is becoming increasingly clear that nanoparticle exposure, among other things, may cause oxidative stress, which could potentially account for many of the adverse effects of nanoparticles (Fig. 5.4 [41]).

Toxicity, or lack of it, of C_{60} fullerenes has been the subject of vigorous debate. Whereas C_{60} has been reported to be extremely toxic to a variety of cultured cells due to its ability to cause peroxidation of membrane lipids, other studies demonstrated that culturing cells with C_{60} does not cause detectable cytotoxic effects [42]. Cells cultured in presence of several hundred micrograms of C_{60} did not show abnormalities in cell proliferation (Fig. 5.5), cell cycle distribution, cell spreading, and cytoskeletal reorganization. Recent studies suggest that the reported cytotoxic

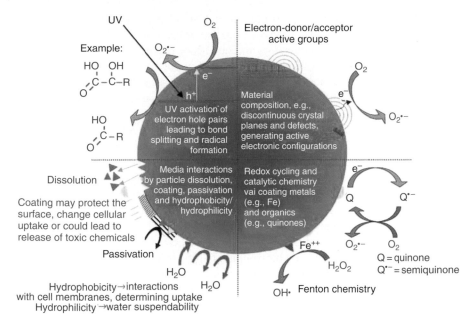

Fig. 5.4 Possible interactions of nanomaterials with cells (reprinted from [23])

effects of C_{60} might be attributed to the solvents (particularly, tetrahydrofuran) used to prepare C_{60}. Further, studies using animal models have also substantiated that C_{60} might not be toxic. Whereas these studies strongly suggest that C_{60} is not toxic, further studies are clearly necessary in this regard.

Studies performed with other nanomaterials (such as nanotubes) are less controversial. Nanotubes have been reported to cause pulmonary and dermal toxicity

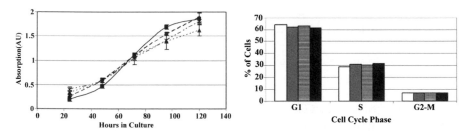

Fig. 5.5 Co-culturing MDA MB 231 breast cancer cells with un-modified C60 do not impact cell proliferation (*left panel*) or cell cycle progression [42]. MCF 10A cells were cultured either in the absence or presence of methanol C_{60} and cell proliferation was assayed by crystal violet staining. ♦ Control, no C_{60}, ■ 10 μg C_{60}, ▲ 50 μg C_{60}, X 250 μg C_{60}. The absorbance of extracted dye is directly proportional to cell growth. Cell cycle analyses right panel were carried out by flow cytometry. MDA MB 231 cells were either untreated (open box) cultured with varying amounts 10 (gray □) 50 (patterned □) and 100 μg (filled ■) of C_{60} for 48 hrs and analyzed for cell cycle progression by flow cytometry

(reviewed in [13]). Pulmonary instillation of nanotubes in rodents is reported to result in granuloma formation, whereas exposure of keratinocytes is known to cause oxidative stress and proinflammatory effects. More recent studies [43] have suggested that surface properties and the presence of impurities (such as metals in nanotubes) are important determinants of cytotoxic responses and showed that SWNTs induce cell death in fibroblasts, possibly through disruption of the cytoskeleton and interference with cell adhesion.

Because of the advantages that quantum dots offer as biological probes, their effects on cell physiology have been examined. While assessing the effects of quantum dots, it is important to consider the composition, size, charge, and the nature of coating material. Further, oxidative, photolytic, and mechanical properties of quantum dots as well as the dosage are key determinants of the biological consequences due to quantum dot exposure [44]. It is also possible that quantum dots may be degraded intracellularly and release toxic metals and capping materials. In contrast, iron oxide particles have not been known to be associated with sub-acute or acute toxicity and are considered to be biocompatible [45, 46].

It is, however, important to recognize that the quantities of C_{60} or other engineered non-biodegradable nanomaterials used in these cytotoxicity studies are substantially higher and are likely to differ from the amounts contained in future nanomedicines, which would be administered at much lower amounts than those used for toxicity studies. For example, the molar quantity of nanomaterials may constitute a relatively small portion of nanomedicines, as the biologically active agents (such as drugs) would also be components of the nanoconjugates, which are anticipated to be the bioactive material. Nanomedicines will be administered at therapeutic and diagnostic doses, but not at the higher amounts used in toxicological studies. Hence, the exposure of nanoparticles to tissues will be significantly less from biomedical uses. Nevertheless, evaluation of potential cytotoxic effects of nanomaterials is important and relevant as accidental, occupational, or chronic exposure from the environment remains a possibility.

5.3.3 Detection of Nanomaterials

It is important to correlate the biological/cytotoxic effects of nanoparticles with the actual intracellular and in vivo quantities that persist upon administration. Whereas a number of techniques, such as electron microscopy or atomic force microscopy, are used to detect nanoparticles in vitro, these techniques may not be suitable or convenient for the routine detection of nanoparticles and their load in tissues. While nanoparticles labeled with radioactivity, fluorescent probes, or chemicals (including drugs, enzymes, biotin, or other reagents) may be readily monitored, it is often difficult to detect low concentrations of unmodified nanoparticles or the metabolites of the conjugated particles. It is also important to consider whether in vivo interactions of nanoparticles (or their metabolites) with serum proteins or tissues would alter their physicochemical properties, thus, rendering their detection more difficult. Emerging research points to some success in this direction. For example, recent work demonstrates that the intrinsic nIR fluorescence may be useful to detect and follow the pharmacokinetics of SWNTs [47]. Similarly HPLC detection of nanogram quantities of C_{60} fullerenes from protein containing samples has been reported [48]. In addition, this method might be applicable for the detection of C_{60} in plasma and skin samples.

5.3.4 Uptake and Clearance of Nanoparticles

The successful application of nanoparticles for diagnostic and therapeutic purposes is dependent on a number of cellular and physiological factors which have been established for conventional drugs. Some of the desired characteristics of nanomedicines include: slower clearance from circulation, improved bioavailability, stability, uptake by target tissue, and entry into the appropriate subcellular compartment. These factors are elegantly summarized by Lacerda et al. (Fig. 5.6; [7]).

Several laboratories have shown the presence of intracellular nanoparticles. Although detailed mechanisms of cellular uptake and intracellular fate of the nanoparticles remain unknown, functionalized SWNTs [49] and gold nanoparticles [50] appear to enter the cells through receptor-mediated endocytic pathways. Thus,

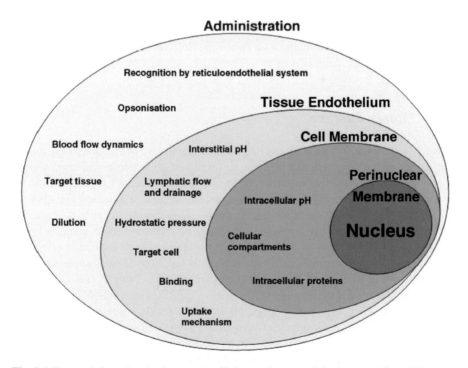

Fig. 5.6 Factors influencing the therapeutic efficiency of nanoparticles (reprinted from [7])

it is generally accepted that particles <100 nm are taken up by endocytic pathways and larger particles are phagocytosed [51].

For any drug to achieve the desirable therapeutic effect, it is necessary to remain in the circulation long enough and reach the required concentration to exert its pharmacological effects. However, particulate drugs are rapidly removed from circulation by mononuclear phagocytic systems, also known as the reticulo endothelial system (RES) [51]. Nanoparticles are rapidly coated with serum proteins in a process known as opsonization and are ingested by phagocytes via endocytosis. If the particle is biodegradable, the phagocytes break them down via lytic mechanisms. However, non-biodegradable particles such as engineered nanoparticles depending on their size are cleared through renal systems or sequestered in the liver or spleen. Hydrophobic particles are more rapidly opsonized than hydrophilic particles as serum proteins show enhanced absorbability to the former type of particles. Therefore, to evade or slow the opsonization process, particles are modified by surface adsorption or grafting hydrophilic compounds such as PEG. Consistent with these general principles, earlier studies have shown that PEGylated C_{60} fullerenes are accumulated in liver [16], whereas functionalized nanotubes are excreted in urine [52].

5.4 Current and Potential Applications of Nanoparticles

From the foregoing discussion it is clear that a number of factors will influence the use of nanoparticles for biomedical applications. Here, specific uses of nanoparticles for the treatment and diagnosis of two human diseases will be considered.

5.4.1 Cancer Therapy

In most cases, cancer-related deaths occur due to the failure of chemotherapy and/or radiation therapy of the metastatic disease. Therefore, for a successful cancer treatment, it is critical to detect tumors early on during the disease progression, detect and ablate tumor metastasis. Nanotechnology has several applications in improving cancer therapy and some nanosized drugs are currently in clinical trials [2]. Prominent among those are liposomal doxorubicin and albumin conjugated Paclitaxel, which have been shown to reduce toxicity due to the adverse side effects of respective drugs. Liposomal doxorubicin (Doxil) has been shown to be an effective anti-neoplastic agent with improved biodistribution (longer plasma circulation times) which reduces severe dose-limiting cardiotoxicity associated with the drug treatment [53]. Like many drugs, Paclitaxel is poorly soluble in water and is administered as a formulation with Cremophor EL (polyethoxylated castor oil) and causes side effects such as hypersensitivity and nephrotoxicity and neurotoxicity. While several different formulations have been made to minimize the toxic effects of this drug, albumin-conjugated nano-sized paclitaxel (Abraxane), a Cremophor free-formulation, has been shown to be well-tolerated and yet more effective than the conventional drug [54]. Additionally, a number of nanosized formulations of other chemotherapeutic drugs such as 5-fluorouracil and camptothecin are being tested. Further, new formulations of aqueous compatible nanosized drugs (such as paclitaxel) are being developed through supercritical fluid technologies [55].

In other applications, the unique properties of nanoparticles are being exploited for the treatment of cancer. In this strategy, nanoparticles serve as multifunctional therapeutic agents, instead of being utilized as simple passive carriers of cargo. These multifunctional nanoparticles can potentially carry drugs to the target tissue, image the target tissues, and release the cargo in response to a signal or upon reaching appropriate cellular compartment. Further, the targeted nanoparticles may be used as therapeutic agents. Targeting the nanotherapeutic may be achieved by coupling a specific antibody or a small molecular weight ligand (e.g., folic acid) that recognizes a protein selectively expressed on tumor cells [2]. Magnetic iron oxide particles [37], nanotubes [29], or other particles may be used as core nanoparticles, and the imaging can be accomplished through MRI (with iron oxide nanoparticles) or fluorescence methods if quantum dots [23] are incorporated into the multifunctional nanoparticle. For example, optical stimulation with an nIR laser of folic acid receptor targeted SWNTs has been shown to act as a trigger for the release of the internalized nanotube attached cargo (in this case oligodeoxy nucleotides) [32]. Further, nIR was also used to generate local heating and then cause photothermal ablation of the targeted cell populations. In another example, tumor-targeting antibody, Herceptin, which is prepared against HER2 receptor (amplified in breast tumors), was coupled to gold nanoshells. This enabled the gold nanoshells to be targeted to HER2 overexpressing tumor cells and when irradiated with nIR, tumor cells were specifically killed [56]. Further, the ability of fullerenes to generate highly reactive singlet oxygen has also been tested as a potential photodynamic therapy of tumor cells [16]. Thus, the concepts of multifunctional nanoparticles for cancer therapy have been validated in experimental systems.

5.4.2 Diabetes

Diabetes mellitus is a serious metabolic disease in which patients are unable to regulate blood glucose levels. Management of both Type I (insulin-dependent) and Type II (insulin-independent) diseases requires a careful maintenance of blood glucose levels. Type I patients require insulin administration whereas the Type II disease is managed by drugs which reduce glucose and insulin. Nanotechnology-based solutions are also being explored in the management of diabetes. Nanoparticle-based biomolecular sensors have been developed to measure glucose levels accurately in blood. For example, as discussed above, glucose oxidase coupled to nanotubes have been shown to serve as catalytic biomolecular sensors [31]. Several types of biodegradable nanoparticles have been developed as drug carriers for oral [57] and transdermal delivery [58] of insulin as an alternative to insulin injections.

Biocompatible and biodegradable pH-sensitive alginate nanospheres which release insulin for extended periods in less acidic intestinal environments, but not in acidic gastric environments, are being developed [57].

5.5 Summary

Nanomaterials hold immense promise for significantly improving existing diagnosis, therapy and designing novel approaches to treat a variety of human ailments (discussed here was specifically cancer and diabetes). While some of the applications of nanotechnology have been translated into clinical settings, many more potential uses of nanomedicines have been demonstrated in experimental systems. Since a variety of materials can be nanosized, the scope of nanomedicine is also large and may even become larger. At the same time, the impact of nanomaterials on cellular and animal models will need to be carefully evaluated under both acute and chronic exposures at toxicological and pharmacological doses. As this chapter indicated, much more work is needed, especially in nanotoxicity, for this field to advance. It is extremely important to evaluate basic issues (such as the fate of nanomaterials in biological systems, and how the cells and tissues react to the exposure of nanomaterials) in developing nanomedicines. This is the next frontier.

Acknowledgments I thank Dr. Linda Knight, Professor of Diagnostic Imaging at Temple University School of Medicine, for critical reading and helpful suggestions with the manuscript.

References

1. Bosi S, Da Ros T, Spalluto G, Prato, M (2003) Fullerene derivatives: an attractive tool for biological applications. Eur J Med Chem 38:913–23.
2. Brannon-Peppas L, Blanchette, JO (2004) Nanoparticle and targeted systems for cancer therapy. *Adv Drug Deliv Rev* 56:1649–1659.
3. Corot C, Robert P, Idee J-M, Port M (2006) Recent advances in iron oxide nanocrystal technology for medical imaging. Adv Drug Deliv Rev 58:1471–1504.
4. Emerich DF, Thanos CG (2006) The pinpoint promise of nanoparticle-based drug delivery and molecular diagnosis. Biomol Eng 23:171–184.
5. Gao X, Cui Y, Levenson RM, Chung LW, Nie S (2004) In vivo cancer targeting and imaging with semiconductor quantum dots. Nat Biotechnol 22:969–976.
6. Harrison BS, Atala A (2007) Carbon nanotube applications for tissue engineering. Biomaterials 28:344–353.
7. Lacerda L, Bianco A, Prato M, Kostarelos, K (2006) Carbon nanotubes as nanomedicines: from toxicology to pharmacology. Adv Drug Deliv Rev 58:1460–1470.
8. Moghimi SM, Hunter AC, Murray JC (2005) Nanomedicine: current status and future prospects. FASEB J 19:311–330.
9. Nishiyama N, Kataoka K (2006) Current state, achievements, and future prospects of polymeric micelles as nanocarriers for drug and gene delivery. Pharmacol Ther 112:630–648.
10. Sharma P, Brown S, Walter G, Santra S, Moudgil B (2006) Nanoparticles for bioimaging. Adv Colloid Interface Sci 123–126:471–485.
11. Freitas RA Jr (1999) Nanomedicine, volume I: basic capabilities, 1st edn. Landes Bioscience, Georgetown, TX.

12. Freitas RA Jr (2003) Nanomedicine, voulme iia: biocompatibility, 1st edn. Landes Bioscience, Georgetown, TX.

13. Oberdorster G, Oberdorster E, Oberdorster J (2005) Nanotoxicology: an emerging discipline evolving from studies of ultrafine particles. Environ Health Persp 113:823–839.

14. Jensen AW, Maru BS, Zhang X, Mohanty DK, Fahlman BD, Tomalia DA (2005) Preparation of fullerene-shell dendrimer-core nanoconjugates. Nano Lett 5:1171–1173.

15. Dresselhaus MS, Dresselhaus G, Eklund, PC (1996) Science of fullerenes and carbon nanotubes. Academic Press, San Diego, CA.

16. Tabata Y, Murakami Y, Ikada Y (1997) Antitumor effect of poly(ethylene glycol)modified fullerene. Fullerene Sci Technol 5:989–1007.

17. Tsao N, Kanakamma PP, Luh T-Y, Chou C-K, Lei H-Y (1999) Inhibition of escherichia coli-induced meningitis by carboxyfullerence. Antimicrob Agents Chemother 43:2273–2277.

18. Yamago S, Tokuyama H, Nakamura E, Kikuchi K, Kananishi S, Sueki K, Nakahara H, Enomoto S, Ambe F (1995) In vivo biological behavior of a water-miscible fullerene: 14C labeling, absorption, distribution, excretion and acute toxicity. Chem Biol 2:385–389.

19. Ali SS, Hardt JI, Quick KL, Kim-Han JS, Erlanger BF, Huang TT, Epstein CJ, Dugan LL (2004) A biologically effective fullerene (C60) derivative with superoxide dismutase mimetic properties. Free Radic Biol Med 37:1191–1202.

20. Dugan LL, Turetsky DM, Du C, Lobner D, Wheeler M, Almli CR, Shen CK-F, Luh T-Y, Choi DW, Lin T-S (1997) Carboxyfullerenes as neuroprotective agents. PNAS 94:9434–9439.

21. Fumelli C, Marconi A, Salvioli S, Straface E, Malorni W, Offidani AM, Pellicciari R, Schettini G, Giannetti A, Monti D, Franceschi C, Pincelli C (2000) Carboxyfullerenes protect human keratinocytes from ultraviolet-B-induced apoptosis. J Invest Dermatol 115:835–841.

22. Gharbi N, Pressac M, Hadchouel M, Szwarc H, Wilson SR, Moussa F (2005) [60] fullerene is a powerful antioxidant in vivo with no acute or subacute toxicity. *Nano Lett* 5, 2578–2585.

23. Yezhelyev MV, Gao X, Xing Y, Al-Hajj A, Nie S, O'Regan RM (2006) Emerging use of nanoparticles in diagnosis and treatment of breast cancer. The Lancet Oncol 7:657–667.

24. Tagmatarchis N, Shinohara H (2001) Fullerenes in medicinal chemistry and their biological applications. Mini Rev Med Chem 1:339–348.

25. Zakharian TY, Seryshev A, Sitharaman B, Gilbert BE, Knight V, Wilson LJ (2005) A fullerene-paclitaxel chemotherapeutic: synthesis, characterization, and study of biological activity in tissue culture. J Am Chem Soc 127:12508–12509.

26. Toth E, Bolskar RD, Borel A, Gonzalez G, Helm L, Merbach AE, Sitharaman B, Wilson LJ (2005) Water-soluble gadofullerenes: toward high-relaxivity, pH-responsive MRI contrast agents. J Am Chem Soc 127:799–805.

27. Katz E, Willner I (2004) Biomolecule-functionalized carbon nanotubes: applications in nanobioelectronics. Chem phys chem 5:1084–1104.

28. Bianco A, Hoebeke J, Kostarelos K, Prato M, Partidos CD (2005) Carbon nanotubes: on the road to deliver. Curr Drug Deliv 2:253–259.

29. Klumpp C, Kostarelos K, Prato M, Bianco A (2006) Functionalized carbon nanotubes as emerging nanovectors for the delivery of therapeutics. Biochim Biophys Acta 1758: 404–412.

30. Ziegler KJ (2005) Developing implantable optical biosensors. Trends Biotechnol 23:440–444.

31. Barone PW, Baik S, Heller DA, Strano MS (2005) Near-infrared optical sensors based on single-walled carbon nanotubes Nat Mater 4:86–92.

32. Shi Kam NW, O'Connell M, Wisdom JA, Dai H (2005) Carbon nanotubes as multifunctional biological transporters and near-infrared agents for selective cancer cell destruction. PNAS 102: 11600–11605.

33. Alivisatos AP, Gu W, Larabell C (2005) Quantum dots as cellular probes. Annu Rev Biomed Eng 7:55–76.

34. Courty S, Bouzigues C, Luccardini C, Ehrensperger M, Bonneau S, Dahan M, James I (2006) Tracking individual proteins in living cells using single quantum dot imaging. In "Methods in Enzymology", Academic Press.

35. Fountaine TJ, Wincovitch SM, Geho DH, Garfield SH, Pittaluga, S (2006) Multispectral imaging of clinically relevant cellular targets in tonsil and lymphoid tissue using semiconductor quantum dots. Mod Pathol 19:1181–1191.
36. Pittet MJ, Swirski FK, Reynolds F, Josephson L, Weissleder R (2006) Labeling of immune cells for in vivo imaging using magnetofluorescent nanoparticles. Nat Protocols 1: 73–79.
37. Gupta AK, Gupta M (2005) Synthesis and surface engineering of iron oxide nanoparticles for biomedical applications. *Biomaterials* 26:3995–4021.
38. Torchilin VP (2005) Recent advances with liposomes as pharmaceutical carriers. Nat Rev Drug Discov 4:145–160.
39. Hoet PH, Bruske-Hohlfeld I, Salata OV (2004) Nanoparticles – known and unknown health risks. J Nanobiotechnol 2:12.
40. Tsuji JS, Maynard AD, Howard PC, James JT, Lam CW, Warheit DB, Santamaria AB (2006) Research strategies for safety evaluation of nanomaterials, Part IV: risk assessment of nanoparticles. Toxicol Sci 89:42–50.
41. Nel A, Xia T, Madler L, Li N (2006) Toxic potential of materials at the nanolevel. Science 311:622–627.
42. Levi N, Hantgan R, Lively M, Carroll D, Prasad G (2006) C60-Fullerenes: detection of intracellular photoluminescence and lack of cytotoxic effects. *J Nanobiotechnol* 4:14.
43. Tian F, Cui D, Schwarz H, Estrada GG, Kobayashi H (2006) Cytotoxicity of single-wall carbon nanotubes on human fibroblasts. Toxicol In Vitro 20:1202–1212.
44. Hardman R (2006) A toxicologic review of quantum dots: toxicity depends on physicochemical and environmental factors. Environ Health Perspect 114:165–172.
45. Sun R, Dittrich J, Le-Huu M, Mueller MM, Bedke J, Kartenbeck J, Lehmann WD, Krueger R, Bock M, Huss R, Seliger C, Grone HJ, Misselwitz B, Semmler W, Kiessling F (2005) Physical and biological characterization of superparamagnetic iron oxide- and ultrasmall superparamagnetic iron oxide-labeled cells: a comparison. Invest Radiol 40:504–513.
46. Weissleder R, Stark DD, Engelstad BL, Bacon BR, Compton CC, White DL, Jacobs P, Lewis J (1989) Superparamagnetic iron oxide: pharmacokinetics and toxicity. AJR Am J Roentgenol 152:167–173.
47. Cherukuri P, Gannon CJ, Leeuw TK, Schmidt HK, Smalley RE, Curley SA, Weisman, RB (2006) Mammalian pharmacokinetics of carbon nanotubes using intrinsic near-infrared fluorescence. Proc Natl Acad Sci USA 103: 18882–18886.
48. Xia XR, Monteiro-Riviere NA, Riviere JE (2006) Trace analysis of fullerenes in biological samples by simplified liquid-liquid extraction and high-performance liquid chromatography. J Chromatogr A 1129:216–222.
49. Shi Kam NW, Jessop TC, Wender PA, Dai H. (2004) Nanotube molecular transporters: internalization of carbon nanotube-protein conjugates into Mammalian cells. J Am Chem Soc 126:6850–6851.
50. Chithrani BD, Ghazani AA, Chan WCW (2006) Determining the size and shape dependence of gold nanoparticle uptake into mammalian cells. Nano Lett 6:662–668.
51. Garnett MC, Kallinteri P (2006) Nanomedicines and nanotoxicology: some physiological principles. Occup Med (Lond) 56:307–311.
52. Singh R, Pantarotto D, Lacerda L, Pastorin G, Klumpp C, Prato M, Bianco A, Kostarelos K (2006) Tissue biodistribution and blood clearance rates of intravenously administered carbon nanotube radiotracers. Proc Natl Acad Sci USA 103:3357–3362.
53. O'Brien MER, Wigler N, Inbar M, Rosso R, Grischke E, Santoro A, Catane R, Kieback DG, Tomczak P, Ackland SP, Orlandi F, Mellars L, Alland L, Tendler C (2004) Reduced cardiotoxicity and comparable efficacy in a phase III trial of pegylated liposomal doxorubicin HCl (CAELYXTM/Doxil(R)) versus conventional doxorubicin for first-line treatment of metastatic breast cancer. Ann Oncol 15:440–449.
54. Moreno-Aspitia A, Perez EA (2005) Nanoparticle albumin-bound paclitaxel (ABI-007): a newer taxane alternative in breast cancer. Future Oncol 1:755–762.

55. Pathak P, Prasad GL, Meziani MJ, Joudeh AA, Sun, YP (2007) Nanosized paclitaxel parti-
    cles from supercritical carbon dioxide processing and their biological evaluation. *Langmuir*
    23:2674–2679.
56. Loo C, Lin A, Hirsch L, Lee MH, Barton J, Halas N, West J, Drezek R (2004) Nanoshell-
    enabled photonics-based imaging and therapy of cancer. Technol Cancer Res Treat 3:33–40.
57. Raj NKK, Sharma CP (2003) Oral insulin – a perspective. J Biomater Appl 17:183–196.
58. Higaki M, Kameyama M, Udagawa M, Ueno Y, Yamaguchi Y, Igarashi R, Ishihara T,
    Mizushima Y (2006) Transdermal delivery of $CaCO_3$-nanoparticles containing insulin. Dia-
    betes Technol Ther 8:369–374.

Chapter 6
Unexpected Reactions by In Vivo Applications of PEGylated Liposomes

Tatsuhiro Ishida and Hiroshi Kiwada

Abstract PEGylated liposome, a liposome coated with polyethylene glycol (PEG), is understood to be biologically inert and, therefore, a suitable vehicle for in vivo applications. The most successful example is the doxorubicin-containing PEGylated liposome, known under the commercial name Doxil/Caelyx for cancer therapy. However, several researchers have found evidence that unexpected immune responses occur even to such polymer-coated liposomes after intravenous injection not only in animals but also in humans. An understanding of the immunological and pathological factors that control the pharmacokinetic and biological behavior of PEGylated liposomes is crucial for the design and safety of a system with optimal therapeutic and/or diagnostic performance. In this study, the interaction of PEGylated liposomes within the biological milieu following parenteral administration is discussed.

Contents

H. Kiwada (✉)
Department of Pharmacokinetics and Biopharmaceutics, Subdivision of Biopharmaceutical Sciences, Institute of Health Biosciences, The University of Tokushima, 1-78-1, Sho-machi, Tokushima 770-8505, Japan
e-mail: hkiwada@ph.tokushima-u.ac.jp

T.J. Webster (ed.), *Safety of Nanoparticles*, Nanostructure Science and Technology,
DOI 10.1007/978-0-387-78608-7_6, © Springer Science+Business Media, LLC 2009

6.1 A Long-Circulating Liposome, PEGylated Liposome

The rapid elimination of liposomes from blood circulation, along with its predominant uptake by the liver and spleen, frustrated early attempts to deliver liposomal drugs to tissues outside the mononuclear phagocyte system (MPS). Due to its unique physical properties, polyethylene glycol (PEG) is commonly used to improve the stability and biological performance of colloidal drug carriers such as liposomes. The typical structure of a PEGylated liposome is described in Fig. 6.1. The grafting of PEG to the surface of liposomes has been clearly shown to extend the circulation lifetime of the liposome. Compared to their counterparts without PEG, these formulations showed extended circulation times in all mammalian species investigated, including mice, rats, dogs, and humans (Blume and Cevc 1990; Klibanov et al. 1990; Allen et al. 1991, 2002; Maruyama et al. 1992; Woodle and Lasic 1992; Woodle 1998). For PEGylated liposomes in the size range of ≤100 nm in diameter, circulation half-lives of 16–24 h have been described in rodents, and as high as 45 h in humans (Allen et al. 1991; Gabizon et al. 1994; Northfelt et al. 1996). PEG's ability to fulfill this role has been attributed mostly to its physical properties such as unlimited water solubility, large excluded volume and high degree of conformational entropy (Lee et al. 1995; Elbert and Hubbell 1996). To this point, the link between the unique physical properties of the polymer and extended circulation lifetimes has been largely cited as the reason behind the reduction or prevention of

Fig. 6.1 A typical structure of a PEGylated liposome. The inner circle is the liposome while the outer blue curved lines represent the PEG functionalization

80–120 nm

protein adsorption to the liposomal surface (Needham et al. 1992; Woodle and Lasic 1992; Torchilin et al. 1994).

6.2 Passive Targeting of PEGylated Liposomes

An advantage of increased circulation times of liposomes is the increased opportunity for selective localization of liposomes to diseased tissues such as solid tumors and regions of infections (Jain 1987; Papahadjopoulos et al. 1991; Bakker-Woudenberg et al. 1992; Vaage et al. 1993, 1997; Wu et al. 1993; Gabizon et al. 1994; Forssen et al. 1996). This process is known as "passive targeting," and it is attainable due to the permeability difference in normal vs. tumor or inflammatory vascular endothelium. In most tissues, the vascular endothelial cells that line the capillaries form tight junctions and are, therefore, relatively impermeable. However, in the inflammatory tissues, the permeability of the vascular endothelium increases through the action of inflammatory mediators. In this case, gaps between endothelium cells occur that are sufficiently large in diameter to allow liposomes to extravasate into tissue interstitium (Lum and Malik 1994). As solid tumors grow, they require additional blood vessels to supply oxygen and nutrients. As new capillaries sprout into the growing tumor, in a process called angiogenesis, they are even more "leaky" (Jain 1987). In the angiogenetic areas, the endothelial fenestrations can reach up to 800 nm in diameter (Yuan et al. 1995). Long-circulating liposomes extravasate through leaky capillaries into the tumor interstitium in greater quantities than in regions with normal vasculature. As a result, concentrations of liposomal anti-cancer drugs or anti inflammation drugs can be increased in such regions several-fold compared to those observed with the free drug.

6.3 Clinical Applications of PEGylated Liposomes

PEGylated liposomes have been under clinical investigations for many years. The only commercially available PEGylated liposome formulation is Doxil/Caelyx, a formulation of PEGylated liposomes that contain DXR (Fig. 6.1). The pharmacology of PEGylated liposomal DXR gives rise to a compound with major advantages that could potentially improve therapeutic response while decreasing toxicity. The encapsulation of DXR into PEGylated liposomes results in a longer circulation time in the body compared with the conventional formulation (Gabizon 2001; Sharpe et al. 2002). The first- and second-phase half-lives of PEGylated liposomal DXR are approximately 5 and 55 h, respectively, compared with approximately 10 min and 30 h, respectively, for conventional DXR (Hussein et al. 2002). As a result, the PEGylated liposomal DXR formulation can be administered at a lower dose than the conventional formulation, potentially reducing the incidence of anthracycline-induced toxicities such as nausea, vomiting, alopecia, myelosuppression, and cardiac toxicity (Safra et al. 2000; Hussein et al. 2002; Safra 2003; O'Brien et al. 2004). The lower toxicity is also related to the encapsulation of DXR into microscopic liposomes. In addition, the PEGylated liposomal DXR preferentially penetrates via angiogenic vasculature with increased microvascular permeability and accumulates in tumor tissue. The increased exposure of tumor cells to DXR released slowly from the accumulated liposomes has the potential of increasing tumor cell killing capacity, theoretically resulting in improved response rates. The longer half-life of the DXR also allows for shortened infusion times and decreased doses, compared with the lengthy infusion times associated with conventional DXR (Hussein et al. 2002). Initially, the indication of this formulation was the treatment of AIDS-related Kaposi's sarcoma (Grunaug et al. 1998; Krown et al. 2004). Recent studies have reported the use of Doxil for the treatment of several other malignant diseases like non-small cell lung cancer, breast cancer, ovarian cancer, prostate cancer, endometrial cancer, hepatocellular carcinoma, and soft-tissue sarcoma (Muggia et al. 1997; Koukourakis et al. 1999; Chidiac et al. 2000; Halm et al. 2000; Hubert et al. 2000; Lyass et al. 2000; Markman et al. 2000; Escobar et al. 2003; Orditura et al. 2004; Coleman et al. 2006). Most recently, several combination regimens incorporating Doxil/Caelyx and free forms of anticancer drugs, such as capecitabine, carboplatin, cyclophosphamide, gemcitabine, mitomycin C, paclitaxel, topotecan, temozolomide, and vinorelbine, are continuously being evaluated in clinical trials (Katsaros et al. 2005; Gnad-Vogt et al. 2005; Mirchandani et al. 2005; Overmoyer et al. 2005; Poh et al. 2005; Bourgeois et al. 2006; Caraglia et al. 2006; Verhaar-Langereis et al. 2006).

PEGylated liposomes have also been used as a carrier for radiolabels to aid in the display of pathological processes scintigraphically (Laverman et al. 1999; Oku 1999; Boerman et al. 2000). For the last three decades, various approaches have been developed to radiolabel PEGylated liposomes with gamma-emitting radionuclides such as Ga-67, In-111, and Tc-99m. A variety of preclinical studies have demonstrated the potential of PEGylated liposomes to image tumor or infectious and inflammatory foci. In a clinical study with Tc-99m-HMPAO-PEGylated liposomes (Dams et al. 2000b), liposome scintigraphy showed high sensitivity (94%) and specificity (89%) in 35 patients who were suspected of having infectious or inflammatory disease. But adverse effects (flushing and chest tightness) were observed in one patient, though both symptoms quickly disappeared by lowering the infusion rate. Then, the same group initiated a study in patients suspected of having an exacerbation of Crohn's disease (Brouwers et al. 2000). While inflamed colon segments were successfully shown in 7 patients, only a moderate relation between the scan grading and the conventional verification procedures (endoscopy or radioscopy) was found. Unfortunately, the clinical study was terminated because of unacceptable adverse effects (tightness in the chest and/or stomach region, mild hyperventilation, and erythema of the face and upper extremities) in 3 out of 9 patients. These studies indicate that radio-labeled PEGylated liposomes can offer an effective and convenient method for the scintigraphic display of focal infection and inflammation. However, the adverse reactions that were encountered impede the further use of this formulation in patients.

6.4 Accelerated Blood Clearance (ABC) of PEGylated Liposomes upon Repeated Injections

6.4.1 ABC Phenomenon

In the absence of encapsulated or surface coupled proteins, PEGylated liposomes are generally considered to be non-immunogenic (van Rooijen and van Nieuwmegen 1980; Alving 1992; Harding et al. 1997). However, it recently has been demonstrated that upon repeated injections, the circulation time of PEGylated liposomes dramatically decreases while their uptake by the liver increases concomitantly (Dams et al. 2000a; Ishida et al. 2003a, b). This so-called ABC phenomenon was observed in rats, mice, and a Rhesus monkey (Dams et al. 2000a; Laverman et al. 2001; Ishida et al. 2003a, b). It was demonstrated that the extent of the ABC phenomenon, induced following prior injection of PEGylated liposomes, depends on the time interval between the first and second injections (Fig. 6.2) (Ishida et al. 2003b, 2006c). When the second dose was given at Day 3, post-first injection, the elimination rate nearly doubled compared to the control. At 4–7 days post-first injection, a further increase in blood clearance was observed. Beyond 8 days after the first injection, the elimination rate dramatically decreased, but significantly ABC was still observed even at 14 days post-first injection (Fig. 6.2A). The accumulation of the second-dose PEGylated liposomes in the liver gradually increased in relation to the time elapsed after the first injection, reaching a maximum level (7- to 8-fold larger than the control) at 4–6 days post-first injection, and gradually decreasing after Day 10 post-first injection. However, even at Day 14, significantly enhanced liver accumulations (5-fold larger than control) were still observed. Significant accumulations of the PEGylated liposomes were observed in the spleen at Days 3 and 4 post-first injection, although at substantially lower levels than in the liver and hardly different from the control values (Fig. 6.2B). The intrahepatic distribution of the PEGylated liposome was studied with a transmission electron microscopy (TEM). A single injection of PEGylated liposomes showed low accumulations in the liver 2 h after injection (Fig. 6.3A), in accordance with established long-circulating properties of such liposomes. After induction of the ABC phenomenon, however, rapid massive internalization of PEGylated liposomes by the macrophages of the liver (Kupffer cells) was observed at 2 h post-injection (Fig. 6.3B and C). These findings clearly indicate that, under conditions in which the ABC phenomenon is induced, PEGylated liposomes can still react with the biological milieu, despite their established steric stabilization with PEG.

6.4.2 Estimated Underlying Mechanism for the ABC Phenomenon

The mechanism underlying the induction of the so-called ABC phenomenon upon repeated injection has been investigated. Dams et al. (2000a) found that the enhanced blood clearance of the PEGylated liposomes was also observed in rats after transfusion of serum from rats that had received PEGylated liposomes 1 week earlier, indicating

Fig. 6.2 Accelerated blood
clearance and enhanced organ
uptake of a second dose of
PEGylated liposomes. Rats were
pretreated with PEGylated
liposomes (lipid dose,
0.001 μmol/kg). (**A**) Blood
clearance of a second dose of
radio-labeled PEGylated
liposomes (5 μmol/kg).
(**B**) Hepatic and splenic
accumulation at 24 h following
the injection. Each value
represents the mean ± SD (*n*=3).
p values apply to differences
between the control and treated
rats. *p< 0.05, ***p< 0.005,
Cited from Ishida et al. (2005)
with permission from Elsevier

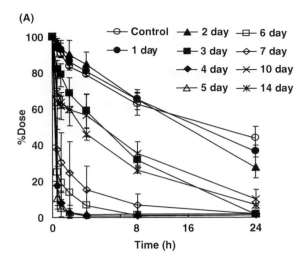

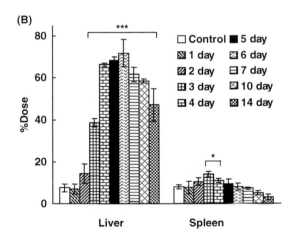

involvement of a soluble serum factor in that process. This observation has been confirmed (Ishida et al. 2003a). In addition, Dams et al. (2000a) showed that the serum
factor is a non-immunoglobulin (IgG or IgM), and a heat-labile, molecule co-eluted
on a size exclusion column with a 150-kDa protein. By contrast, recent studies clearly
showed that upon repeated injection, PEGylated liposomes produced an abundance of
anti-PEG IgM and consequently enhanced blood clearance of subsequently injected
PEGylated liposomes as a result of anti-PEG IgM-mediated complement activation
under certain conditions (Fig. 6.4) (Ishida et al. 2003a, b, 2005, 2006a; Wang et al. in
press). It seems that an increase in the promotion of anti-PEG IgM production could be
the trigger that converts the PEGylated liposome from a non-harmful into a potentially
harmful pathological formulation.

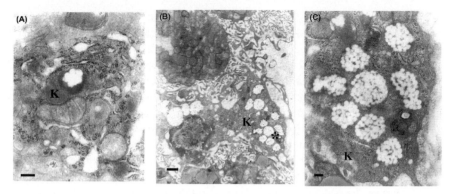

Fig. 6.3 Electron micrographs of rat livers after intravenous administration of 5 μmol/kg PEGylated liposomes as a second dose, 5 days after injection of saline (**A**) or a pre-dose of 0.001 μmol/kg PEGylated liposomes (**B** and **C**). (**A**) Kupffer cell, 2 h after injection of 5 μmol/kg PEGylated liposomes into a rat that had received saline 5 days earlier; bar = 200 nm. (**B** and **C**) Kupffer cells, 2 h after injection of 5 μmol/kg PEGylated liposomes into a rat that had received PEGylated liposomes (0.001 μmol/kg) 5 days earlier; bar in (**B**) = 1 μm. (**C**) higher magnification view of the region marked in (**B**); bar =200 nm. K, Kupffer cell, (Cited from Ishida et al. (2006c) with permission from Elsevier)

Fig. 6.4 Representation of the sequence of events leading from anti-PEG IgM induction to accelerated clearance of PEGylated liposomes. Cited from Ishida et al. (2006c) with permission from Elsevier

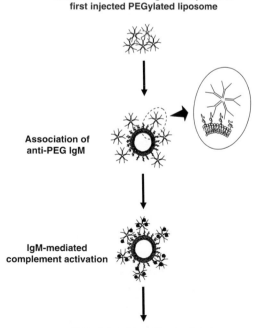

Secretion of anti-PEG IgM in response to first injected PEGylated liposome

Association of anti-PEG IgM

IgM-mediated complement activation

Complement receptor-mediated endocytosis single by liver macrophages

Cheng and co-workers (Cheng et al. 1999; Tsai et al. 2001) demonstrated that a monoclonal antibody (IgM) against PEG obtained following immunization with PEGylated β-glucuronide recognizes the repeating $O–CH_2–CH_2$ subunit (16 units) of PEG. Thus, the anti-PEG IgM produced by intravenous injection of PEGylated liposomes, if it is monoclonal, may bind to the similar epitope of PEG on the liposome. Generally, IgM antibodies tend to be of low affinity. However, IgM molecules form pentamers whose ten antigen-binding sites can bind simultaneously to multivalent antigens. This compensates for the relatively low affinity of the IgM monomers by multipoint binding, conferring high overall avidity. The high avidity, a basic characteristic of IgM, might provide sufficiently strong binding of the generated anti-PEG IgM to the PEGylated liposome, specifically to the PEG moiety, to overcome the protective effect and the free movement of the PEG chains on the liposome surface.

Furthermore, a study revealed that the enhanced blood clearance effect was not induced in splenectomized rats, which goes along with a dramatic reduction of serum IgM level (Ishida et al. 2006b). In addition, this study showed that the immune reaction in the spleen against the PEGylated liposomes extends over a period of at least 2–3 days following the first administration and then anti-PEG IgM is produced (Ishida et al. 2006b). It has been reported that B cells in the spleen are responsible for the first line of defense and are able to produce large amounts of neutralizing antibodies in a short period (3–4 days) (Martin et al. 2001; Zandvoort and Timens 2002). These indicate that the spleen plays an important role in promoting the formation of IgM reactive against PEG. It is well known that the spleen plays a central role in the primary defense against all types of antigens that appear in circulation, and that it is a major site of antibody production (Koch et al. 1982; Bohnsack and Brown 1986). Liposomes with long-circulating properties, such as PEGylated liposomes, reportedly tend to accumulate in the spleen rather than in the liver. The PEG polymer could be characterized by a highly repetitive structure bearing similarity to a class-2 thymus-independent (TI-2) antigen. Therefore, the following hypothesis leading to the production of anti-PEG IgM is proposed. Once the PEGylated liposomes reach the spleen, they bind and crosslink to surface immunoglobulins on PEG (or PEGylated liposome)-reactive B-cells in that organ and consequently trigger the production of an anti-PEG IgM that is independent of T-cell help.

6.4.3 Effect of Encapsulation of Doxorubicin (DXR) on Inducing the ABC Phenomenon

Although PEGylated liposomal doxorubicin (DXR) (Doxil/Caelyx) has already been in clinical use, such phenomenon has not been reported in patients. Consistent with this fact, Charrois and Allen (2003) recently showed that in a murine model, after repeated administration of DXR-loaded PEGylated liposomes, varying both the dosing schedule and the dosing intensity, the pharmacokinetics and biodistribution of the liposomes did not change. As described above, liposomal encapsulation of DXR accounts for several benefits over the free form of the drug. Of foremost importance, liposomal formulations buffer the acute cardiotoxicity of DXR (Safra 2003; O'Brien et al. 2004), while also extending blood circulation time and improving drug accumulation in malignant tissues (Gabizon et al. 1994). It is also known that DXR encapsulation in liposomes can result in increased DXR delivery to phagocytic cells, which, in turn, are killed by the encapsulated drug (Daemen et al. 1995). Laverman et al. (2001) demonstrated that Caelyx did not induce such ABC phenomenon in rats, while empty PEGylated liposomes did. They speculated that a toxic effect of DXR on the hepatosplenic macrophages might have impaired the role of these cells in the induction of the phenomenon, especially the production of unknown non-immunoglobulin serum factor(s). By contrast, it was recently reported that the binding of anti-PEG IgM to PEGylated liposomes and subsequent complement activation on the liposomes are a major cause of the phenomenon (Ishida et al. 2006c; Wang et al. in press). It seemed that the spleen plays an important role in the production of anti-PEG IgM (Ishida et al. 2006b). These have led to the proposal of an alternative to the theory of Laverman et al. (2001): that DXR inside the first dose of injected PEGylated liposomes reduces the production of anti-PEG IgM by interference with the proliferation of B cells in the spleen, and consequently prevents the induction of the phenomenon.

A first injection of PEGylated liposomes containing encapsulated DXR showed no rapid clearance of the second dose of PEGylated liposomes (Ishida et al. 2006a). Furthermore, Western blot analysis (Fig. 6.5A) and quantitative analysis (Fig. 6.5B) revealed abundant binding of IgM to PEGylated liposomes when the liposomes were incubated in serum from rats that had received "empty" PEGylated liposomes. Substantially less binding of IgM was found when the liposomes were incubated in serum from rats treated with DXR-loaded PEGylated liposomes. For both the empty and the DXR-containing liposomes, the amounts of IgM binding to the liposomes decreased with an increasing dose of injected liposomes. The effect of lipid dose on the induction of the phenomenon is described in a later section. Serum obtained from rats following injection of "empty" PEGylated liposomes caused complement activation by the addition of PEGylated liposomes. By contrast, no complement activation was detected with serum from rats that had been treated with DXR-loaded PEGylated liposomes. Based on these findings, it was concluded that DXR released from liposomes accumulating in the spleen impairs the production of PEG-specific IgM as a consequence of the inhibition of B cell-proliferation and/or killing of proliferating B cells, and thus prevents the induction of the enhanced clearance of a second dose of PEGylated liposomes.

Laverman et al. (2001) have proposed that hepatosplenic macrophages are responsible for the induction of the ABC phenomenon. They investigated the pharmacokinetic behavior of test-dose PEGylated liposomes in rats that were depleted of hepatosplenic macrophages by intravenous injection of unsized multilamellar liposomes containing clodronate at the time the first dose of "empty" PEGylated liposomes was administered. Two weeks later, when the macrophage populations had recovered and the second dose of PEGylated liposomes was injected, the long circulation time and low hepatosplenic uptake commonly observed for PEGylated

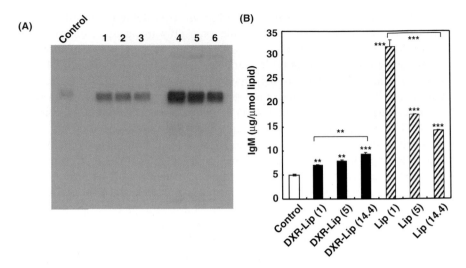

Fig. 6.5 IgM associated with PEGylated liposomes incubated in serum. PEGylated liposomes were incubated with naïve serum or with serum collected 5 days after injection with different doses of PEGylated liposomes with or without encapsulated DXR. (**A**) Immunoblotting: The incubation mixture was subsequently chromatographed on a Sepharose 4 Fast Flow column. The recovered liposome samples were applied to 7% SDS–PAGE gels and electrophoresed under reducing conditions. Then the gels were analyzed by Western blotting with antibodies specific for rat IgM. Control, naïve serum. Lanes 1–6, sera from pre-dosed animals. Lanes 1–3 pre-dosing with DXR-containing PEGylated liposomes. Lanes 4–6, pre-dosing with empty PEGylated liposomes. Lanes 1, 2, 3 and Lanes 4, 5, 6 represent lipid doses of 1, 5, and 14.4 μmol PL/kg, respectively. (**B**) Quantification of IgM: Quantification of IgM in the serum proteins associated with PEGylated liposomes was performed using the rat IgM ELISA Quantification Kit. Captions under the bars have the same meaning as in Fig. 6.1(**B**). The values represent the mean ± SD of three independent experiments. p values apply to differences between naïve serum and liposome-treated serum and between sera from rats that received PEGylated liposomes. **$p<$ 0.01, ***$p<$ 0.005 (cited from Ishida et al. (2006a) with permission from Elsevier)

liposomes were observed. It is important to note, however, that Beringue et al. (2000) reported that, in addition to hepatosplenic macrophages, also most follicular dendritic cells and B lymphocytes, but not T cells, in the spleen are transiently depleted by intravenous injection of clodronate-containing liposomes. It is now believed that the lack of B lymphocytes in the clodronate-treated rats might have been responsible for the lack of induction of the ABC phenomenon in the study of Laverman et al. (2001).

As demonstrated above, the encapsulation of DXR in PEGylated liposomes abrogates the immunogenic response against PEGylated liposomes. Earlier it was shown that encapsulation of DXR in liposomes decreases its dose-limiting toxic side effects such as cardiotoxicity while maintaining, or even increasing, its antitumor effect (Safra 2003; O'Brien et al. 2004). Taken together these observations may indicate that liposomes containing chemotherapeutic agents such as DXR are relatively safe for clinical use even if repeated injections are required. However, Daemen et al. (1997) reported that injection of DXR-loaded PEGylated liposomes has a toxic effect on the liver macrophages both in terms of specific phagocytic activity and cell numbers, although the observed toxicities were less than those induced by DXR-loaded conventional liposomes (without PEGylation). It is known that defects in the phagocytic uptake activity of macrophages enhance metastatic growth of tumors in numerous animal studies (Phillips 1989). Considering these results, injection schemes of DXR-loaded PEGylated liposomes such as Doxil/Caelyx should be designed to permit the recovery of the macrophage population between injections.

6.4.4 Effect of Lipid Dose on Inducing ABC Phenomenon

The present study showed that the induction of the phenomenon by PEGylated liposomes decreased with increases in the first dose (0.001–5 μmol/kg). A strong inverse relationship was found to exist between the dose of initially injected PEGylated liposomes and the extent to which the ABC phenomenon was induced: the higher the dose the smaller the phenomenon (Fig. 6.6) (Ishida et al. 2005). Due to earlier results (Ishida et al. 2006a, b), the lipid dose, less than 1 μmol phospholipids/kg, is likely to be sufficient to activate PEG (or PEGylated liposome)-reactive B-cells in spleen. At the recommended dose intensity of Caelyx for patients (Lyass et al. 2000), a higher lipid dose of PEGylated liposomes (15 μmol phospholipids/kg or even more) is generally administered. Thus, the higher lipid doses (>1 μmol phospholipids/kg) used in this study might lead the B cells to become apoptotic and thus anergic, even though the encapsulated DXR did not work to kill the proliferating B-cells in the spleen. These observations indicate that the higher lipid doses used in the clinic also may present a major cause of the attenuation of the induction phase of the phenomenon.

DXR-loaded PEGylated liposomes appeared to impair the production of anti-PEG IgM in a lipid dose (and thus DXR dose)-dependent manner (Fig. 6.5A) and consequently attenuate the ABC phenomenon (Ishida et al. 2006a). This may reflect the amount of DXR delivered to the liver and spleen in the encapsulated form. It is likely that DXR released from the liposomes in the tissue (taken up by either macrophages or other cells) not only impairs the function of B lymphocytes but also the phagocytic capacity of the macrophages.

The underlying mechanism for causing a strong inverse relationship between the dose of initially injected PEGylated liposomes and the extent to which the ABC phenomenon is induced is still unclear. The so-called second class of thymus-independent antigens (TI-2) likely acts by extensively cross linking the cell-surface immunoglobulins of specific B-cells, resulting in secretion of both IgM and IgG from the B-cells. To achieve sufficient cross linking with epitopes, the epitope density is critical for the activation of B-cells by the TI-2 antigen: at too low a density the level of receptor cross linking is insufficient to activate the cell; at too high a density the cell becomes anergic. The PEG polymer is also characterized by a highly repetitive structure similar to the TI-2 antigens. The low dose of PEGylated liposomes sufficient to induce the ABC phenomenon might represent a sufficiently high epitope (PEG) density to activate the B-cells. The higher lipid doses might lead the B-cells to become apoptotic and thus anergic.

Fig. 6.6 Effect of lipid dose of first injection on the induction of the ABC phenomenon by PEGylated liposomes. Rats were pretreated with PEGylated liposomes at a dose of 0.001, 0.01, 0.1, 1, or 5 μmol/kg of body weight. Rats pretreated with HEPES buffered saline (pH 7.4), instead of PEGylated liposomes, served as controls. At Day 5 post-first injection, radio-labeled PEGylated liposomes (5 μmol/kg) were intravenously injected. (**A**) Blood clearance profile of the radio-labeled PEGylated liposomes. (**B**) Hepatic and splenic accumulation of the radio-labeled PEGylated liposomes at 24 h following the injection. Each value represents the mean ± SD of three separate experiments. p values apply to differences between the control and treated mice. *$p <$ 0.05, ** $p<0.01$, ***$p<0.005$ (cited from Ishida et al. (2005) with permission from Elsevier)

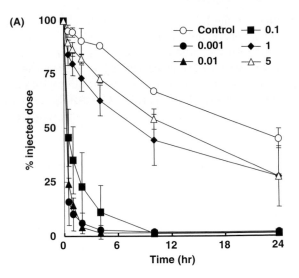

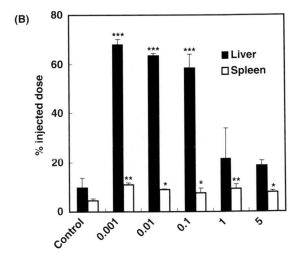

6.5 Adverse Reactions to PEGylated Liposomes Observed in Clinical Use

In many pre-clinical or clinical studies with Doxil/Caelyx, a PEGylated liposomal DXR formulation, acute as well as subacute side effects were encountered during or after the infusion of the liposomal formulation (Uziely et al. 1995; Goebel et al. 1996; Northfelt et al. 1997; Koukourakis et al. 1999; Krown et al. 2004). Most attention has been focused on DXR-related subacute side effects, which include mucositis, stomatitis, and cutaneous toxicity – referred to as hand–foot syndrome (palmar-plantar erythrodysesthesia, PPE) (Lotem et al. 2000; Lyass et al. 2000; Gabizon 2001; Waterhouse et al. 2001; Gordinier et al. 2006). The PPE occurs in up to 13–50% of treated patients, a much greater frequency than is seen with the free form of DXR (Alberts and Garcia 1997; Muggia et al. 1997; Johnston and Gore 2001; Gibbs et al. 2002; Coleman et al. 2006). English et al. (Gordon et al. 2001; English et al. 2003) recently reported the clinical and histopathologic findings of an intertriginous dermatitis due to PEGylated liposomal DXR toxicity and discussed possible mechanisms. The authors showed possibilities that the enhanced delivery of DXR in liposomal form promotes cutaneous side effects of the drug and the PEGylated liposomal form of DXR enhances its compartmentalization in skin while decreasing its degradation, thereby promoting a delay (3–4 weeks following therapy) in its cutaneous effects. The incidence of PPE can be limited by adjusting the injection interval, dosage regimen, and dose (Alberts and Garcia 1997; Gabizon and Muggia 1998; Krown et al. 2004).

6.5.1 Acute Reactions (Hypersensitivity Reaction)

Signs and symptoms of acute reactions consist of shortness of breath, widespread erythemafacil flushing, chest pain, back pain, and changes in blood pressure; all resolve within minutes after termination of the liposome infusions. The symptoms are attenuated by medication (or prevented by premedication) that contain adrenaline and/or hydrocortisone or methylprednisolone (Laing et al. 1994; Uziely et al. 1995; Koukourakis et al. 1999; Huang et al. 2004). Not surprisingly, these acute side effects offered no reason for alarm because the events were minor compared to the usual toxicity of anticancer drugs, and seemed to be balanced by the therapeutic benefits of the drug. On the other hand, if "empty" PEGylated liposomes cause similar reactions when used as a diagnostic agent for the scintigraphic detection of infection and inflammation, such adverse events are unacceptable. Dams et al. (2000b) reported that 2 of 35 patients experienced mild flushing and tightness of the chest during the infusion of Tc-99m-labeled PEGylated liposomes (without encapsulated drugs) for clinical evaluation. Such adverse reactions subsided when the infusion was temporarily stopped.

6.5.2 Influence of Infusion Rate and Regimen on the Acute Reactions (Hypersensitivity Reaction)

It seems that the infusion rate affects the level of acute reactions caused by injection of PEGylated liposomes. For Doxil, the acute side effects could be prevented by reducing the infusion rate or stopping the infusion (Gabizon et al. 1994). However, for Tc-99m-labeled PEGylated liposomes, the reactions still occurred at a slow infusion rate (1 μmol lipid/min) and even with a tenfold-lower lipid dose (5 μmol/kg for Doxil vs. 0.5 μmol/kg for Tc-99m-labeled PEGylated liposomes) (Dams et al.

2000b). This induction appears to indicate that even a minimal amount of PEGylated liposomes could induce such acute side effects.

In all reported cases, subsequent injections could be given without problems regardless of whether these patients experienced side effects during the first infusion. However, Koukourakis et al. (1999) described one patient who experienced the same symptoms as during the first infusion (heavy chest feeling). The fact that even a minimal amount of PEGylated liposomes could induce side effects suggests that the observed reactions could be classified as hypersensitivity or a pseudoallergic reaction (Szebeni et al. 1999, 2000).

6.5.3 The Role of Complement Activation on the Acute Reactions (Hypersensitivity Reaction)

Chanan-Khan et al. (2003) reported that 13 hypersensitivity reactions were observed out of 29 cancer patients treated with Doxil. Plasma complement terminal complex (SC5b-9) levels were elevated relative to the baseline in 80% of reactor patients and 50% of non-reactor patients at 10 min post-infusion, suggesting that Doxil causes complement activation in a larger percentage of cancer patients. The relative extent of complement activation was also significantly greater in the reactor group than in the non-reactor group. Both the incidence of hypersensitivity reaction and the extent of complement activation showed positive correlation with the initial rate of Doxil infusion in the range of 0.1–1.0 mg/min. Brouwers et al. (2000) reported that 2 of 35 patients showed hypersensitivity reaction when radio-labeled (Tc-99m) PEGylated liposomes were injected. In another trial with non-radio-labeled PEGylated liposomes, 3 of 9 patients showed such a reaction. Direct indication for the role of complement activation in the observed hypersensitivity reaction came from later clinical trials. Analysis of serum samples drawn from a patient just prior to the infusion and immediately after the occurrence of the reaction indicated a remarkable decrease in complement factors (C3, C4, and factor B). Whether a hypersensitivity reaction is a direct result of the complement activation or whether the complement consumption is a result of the reaction remains to be established. However, at the least, these results suggested that complement activation may be an important contributing factor in the genesis of hypersensitivity reaction caused by intravenous injection (infusion) of PEGylated liposomes.

The major role of the complement system is to mark invading pathogens, immune complexes, and similar pathogenic materials for removal by phagocytosis (Bradley and Devine 1998). Many studies have described the role of liposomes in complement activation (Patel 1992; Devine and Marjan 1997; Bradley and Devine 1998; Ishida et al. 2001, 2002; Yan et al. 2005) and most of those studies have focused on the role of the complement system in the opsonization of liposomes and consequent clearance of liposomes from circulation. As described above, complement activation by PEGylated liposomes seems to be essential for causing a PEGylated liposome-related hypersensitivity reaction (Szebeni et al. 2000, 2002; Szebeni 2005a, b).

6.6 Conclusions

PEGylated liposomal formulation has been approved and is currently used in clinical settings (Muggia et al. 1997; Grunaug et al. 1998; Koukourakis et al. 1999; Chidiac et al. 2000; Halm et al. 2000; Hubert et al. 2000; Lyass et al. 2000; Markman et al. 2000; Escobar et al. 2003; Krown et al. 2004; Orditura et al. 2004; Coleman et al. 2006). The repeated injection of such liposomal formulations with constant rate infusion is frequently used in treatment protocols (Goebel et al. 1996; Halm et al. 2000; Hubert et al. 2000). The occurrence of such ABC phenomenon could compromise the therapeutic efficacy of drugs associated with liposomes and, if the liposomes contain a toxic drug, the strongly increased uptake in the liver could cause severe liver toxicity (Daemen et al. 1995). Also, acute adverse reactions, including hypersensitivity reaction, caused by intravenous injections of PEGylated liposomes may impede use of the PEGylated formulation in patients, even if the formulations have superior advantages. The degree of caution called for when treating patients with certain regimens may therefore apply beyond the applications of PEGylated formulations, including proteins, peptides, DNAs, ODNs, and siRNAs.

In conclusion, PEGylated liposomes represent a suitable ling-circulating carrier system for in vivo application, but one must consider that PEGylated liposomes are not inert vehicles in vivo. Such reactions may also turn out to be beneficial, if how to control the carrier-induced responses of the immune system is learned. Therefore, the processes involved in the interactions of drug carriers with the biological milieu following in vivo administration need further study.

Acknowledgments We thank Dr. James L. McDonald for his helpful advice in writing the English manuscript.

References

Alberts DS, Garcia DJ (1997) Safety aspects of pegylated liposomal doxorubicin in patients with cancer. Drugs 54:30–35

Allen C, Dos Santos N, Gallagher R, Chiu GN, Shu Y, Li WM, Johnstone SA, Janoff AS, Mayer LD, Webb MS, Bally MB (2002) Controlling the physical behavior and biological performance of liposome formulations through use of surface grafted poly(ethylene glycol). Biosci Rep 22:225–250

Allen TM, Hansen C, Martin F, Redemann C, Yau-Young A (1991) Liposomes containing synthetic lipid derivatives of poly(ethylene glycol) show prolonged circulation half-lives in vivo. Biochim Biophys Acta 1066:29–36

Alving CR (1992) Immunologic aspects of liposomes: presentation and processing of liposomal protein and phospholipid antigens. Biochim Biophys Acta 1113:307–322

Bakker-Woudenberg IA, Lokerse AF, ten Kate MT, Storm G (1992) Enhanced localization of liposomes with prolonged blood circulation time in infected lung tissue. Biochim Biophys Acta 1138:318–326

Beringue V, Demoy M, Lasmezas CI, Gouritin B, Weingarten C, Deslys JP, Andreux JP, Couvreur P, Dormont D (2000) Role of spleen macrophages in the clearance of scrapie agent early in pathogenesis. J Pathol 190:495–502

Blume G, Cevc G (1990) Liposomes for the sustained drug release in vivo. Biochim Biophys Acta 1029:91–97

Boerman OC, Laverman P, Oyen WJ, Corstens FH, Storm G (2000) Radiolabeled liposomes for scintigraphic imaging. Prog Lipid Res 39:461–475

Bohnsack JF, Brown EJ (1986) The role of the spleen in resistance to infection. Annu Rev Med 37:49–59

Bourgeois H, Ferru A, Lortholary A, Delozier T, Boisdron-Celle M, Abadie-Lacourtoisie S, Joly F, Chieze S, Chabrun V, Gamelin E, Tourani JM (2006) Phase I-II study of pegylated liposomal doxorubicin combined with weekly paclitaxel as first-line treatment in patients with metastatic breast cancer. Am J Clin Oncol 29:267–275

Bradley AJ, Devine DV (1998) The complement system in liposome clearance: can complement deposition be inhibited? Adv Drug Deliv Rev 32:19–29

Brouwers AH, De Jong DJ, Dams ET, Oyen WJ, Boerman OC, Laverman P, Naber TH, Storm G, Corstens FH (2000) Tc-99m-PEG-Liposomes for the evaluation of colitis in Crohn's disease. J Drug Target 8:225–233

Caraglia M, Addeo R, Costanzo R, Montella L, Faiola V, Marra M, Abbruzzese A, Palmieri G, Budillon A, Grillone F, Venuta S, Tagliaferri P, Del Prete S (2006) Phase II study of temozolomide plus pegylated liposomal doxorubicin in the treatment of brain metastases from solid tumours. Cancer Chemother Pharmacol 57:34–39

Chanan-Khan A, Szebeni J, Savay S, Liebes L, Rafique NM, Alving CR, Muggia FM (2003) Complement activation following first exposure to pegylated liposomal doxorubicin (Doxil): possible role in hypersensitivity reactions. Ann Oncol 14:1430–1437

Charrois GJ, Allen TM (2003) Multiple injections of pegylated liposomal Doxorubicin: pharmacokinetics and therapeutic activity. J Pharmacol Exp Ther 306:1058–1067

Cheng TL, Wu PY, Wu MF, Chern JW, Roffler SR (1999) Accelerated clearance of polyethylene glycol-modified proteins by anti-polyethylene glycol IgM. Bioconjug Chem 10:520–528

Chidiac T, Budd GT, Pelley R, Sandstrom K, McLain D, Elson P, Crownover R, Marks K, Muschler G, Joyce M, Zehr R, Bukowski R (2000) Phase II trial of liposomal doxorubicin (Doxil) in advanced soft tissue sarcomas. Invest New Drugs 18:253–259

Coleman RE, Biganzoli L, Canney P, Dirix L, Mauriac L, Chollet P, Batter V, Ngalula-Kabanga E, Dittrich C, Piccart M (2006) A randomised phase II study of two different schedules of pegylated liposomal doxorubicin in metastatic breast cancer (EORTC-10993). Eur J Cancer 42:882–887

Daemen T, Hofstede G, Ten Kate MT, Bakker-Woudenberg IA, Scherphof GL (1995) Liposomal doxorubicin-induced toxicity: depletion and impairment of phagocytic activity of liver macrophages. Int J Cancer 61:716–721

Daemen T, Regts J, Meesters M, Ten Kate MT, Bakker-Woudenberg IA, Scherphof GL (1997) Toxicity of doxorubicin entrapped within long-circulating liposomes. J Control Release 44:1–9

Dams ET, Laverman P, Oyen WJ, Storm G, Scherphof GL, van Der Meer JW, Corstens FH, Boerman OC (2000a) Accelerated blood clearance and altered biodistribution of repeated injections of sterically stabilized liposomes. J Pharmacol Exp Ther 292:1071–1079

Dams ET, Oyen WJ, Boerman OC, Storm G, Laverman P, Kok PJ, Buijs WC, Bakker H, van der MeerJW, Corstens FH (2000b) 99mTc-PEG liposomes for the scintigraphic detection of infection and inflammation: clinical evaluation. J Nucl Med 41:622–630

Devine DV, Marjan JM (1997) The role of immunoproteins in the survival of liposomes in the circulation. Crit Rev Ther Drug Carrier Syst 14:105–131

Elbert DL, Hubbell JA (1996) Surface treatments of polymers for biocompatibility. Ann Rev Mat Sci 26:365–394

English JC, 3rd, Toney R, Patterson JW (2003) Intertriginous epidermal dysmaturation from pegylated liposomal doxorubicin. J Cutan Pathol 30:591–595

Escobar PF, Markman M, Zanotti K, Webster K, Belinson J (2003) Phase 2 trial of pegylated liposomal doxorubicin in advanced endometrial cancer. J Cancer Res Clin Oncol 129:651–654

Forssen EA, Male-Brune R, Adler-Moore JP, Lee MJ, Schmidt PG, Krasieva TB, Shimizu S, Tromberg BJ (1996) Fluorescence imaging studies for the disposition of daunorubicin liposomes (DaunoXome) within tumor tissue. Cancer Res 56:2066–2075

Gabizon A (2001) Pegylated liposomal doxorubicin: metamorphosis of an old drug into a new form of chemotherapy. Cancer Invest 19:424–436

Gabizon A, Catane R, Uziely B, Kaufman B, Safra T, Cohen R, Martin F, Huang A, Barenholz Y (1994) Prolonged circulation time and enhanced accumulation in malignant exudates of doxorubicin encapsulated in polyethylene-glycol coated liposomes. Cancer Res 54:987–992

Gabizon A, Muggia FM (1998) Initial clinical evaluation of pegylated liposomal doxorubicin in solid tumors. In: Woodle MC, Storm G (eds) Long circulating liposomes: old drugs, new therapeutics. Springer-Verlag, Berlin, p 165

Gibbs DD, Pyle L, Allen M, Vaughan M, Webb A, Johnston SR, Gore ME (2002) A phase I dose-finding study of a combination of pegylated liposomal doxorubicin (Doxil), carboplatin and paclitaxel in ovarian cancer. Br J Cancer 86:1379–1384

Gnad-Vogt SU, Hofheinz RD, Saussele S, Kreil S, Willer A, Willeke F, Pilz L, Hehlmann R, Hochhaus A (2005) Pegylated liposomal doxorubicin and mitomycin C in combination with infusional 5-fluorouracil and sodium folinic acid in the treatment of advanced gastric cancer: results of a phase II trial. Anticancer Drugs 16:435–440

Goebel FD, Goldstein D, Goos M, Jablonowski H, Stewart JS (1996) Efficacy and safety of Stealth liposomal doxorubicin in AIDS-related Kaposi's sarcoma, The International SL-DOX Study Group. Br J Cancer 73:989–994

Gordinier ME, Dizon DS, Fleming EL, Weitzen S, Schwartz J, Parker LP, Granai CO (2006) Elevated body mass index does not increase the risk of palmar-plantar erythrodysesthesia in patients receiving pegylated liposomal doxorubicin. Gynecol Oncol 103:72–74

Gordon AN, Fleagle JT, Guthrie D, Parkin DE, Gore ME, Lacave AJ (2001) Recurrent epithelial ovarian carcinoma: a randomized phase III study of pegylated liposomal doxorubicin versus topotecan. J Clin Oncol 19:3312–3322

Grunaug M, Bogner JR, Loch O, Goebel FD (1998) Liposomal doxorubicin in pulmonary Kaposi's sarcoma: improved survival as compared to patients without liposomal doxorubicin. Eur J Med Res 3:13–19

Halm U, Etzrodt G, Schiefke I, Schmidt F, Witzigmann H, Mossner J, Berr F (2000) A phase II study of pegylated liposomal doxorubicin for treatment of advanced hepatocellular carcinoma. Ann Oncol 11:113–114

Harding JA, Engbers CM, Newman MS, Goldstein NI, Zalipsky S (1997) Immunogenicity and pharmacokinetic attributes of poly(ethylene glycol)-grafted immunoliposomes. Biochim Biophys Acta 1327:181–192

Huang JY, Wu CH, Shih IH, Lai PC (2004) Complications mimicking lupus flare-up in a uremic patient undergoing pegylated liposomal doxorubicin therapy for cervical cancer. Anticancer Drugs 15:239–241

Hubert A, Lyass O, Pode D, Gabizon A (2000) Doxil (Caelyx): an exploratory study with pharmacokinetics in patients with hormone-refractory prostate cancer. Anticancer Drugs 11: 123–127

Hussein MA, Wood L, Hsi E, Srkalovic G, Karam M, Elson P, Bukowski RM (2002) A Phase II trial of pegylated liposomal doxorubicin, vincristine, and reduced-dose dexamethasone combination therapy in newly diagnosed multiple myeloma patients. Cancer 95:2160–2168

Ishida T, Harashima H, Kiwada H (2001) Interactions of liposomes with cells in vitro and in vivo: opsonins and receptors. Curr Drug Metab 2:397–409

Ishida T, Harashima H, Kiwada H (2002) Liposome clearance. Biosci Rep 22:197–224

Ishida T, Maeda R, Ichihara M, Irimura K, Kiwada H (2003a) Accelerated clearance of PEGylated liposomes in rats after repeated injections. J Control Release 88:35–42

Ishida T, Masuda K, Ichikawa T, Ichihara M, Irimura K, Kiwada H (2003b) Accelerated clearance of a second injection of PEGylated liposomes in mice. Int J Pharm 255:167–174

Ishida T, Harada M, Wang XY, Ichihara M, Irimura K, Kiwada H (2005) Accelerated blood clearance of PEGylated liposomes following preceding liposome injection: effects of lipid dose and PEG surface-density and chain length of the first-dose liposomes. J Control Release 105: 305–317

Ishida T, Atobe K, Wang X, Kiwada H (2006a) Accelerated blood clearance of PEGylated lipo-
    somes upon repeated injections: effect of doxorubicin-encapsulation and high-dose first injec-
    tion. J Control Release 115:251–258
Ishida T, Ichihara M, Wang X, Kiwada H (2006b) Spleen plays an important role in the induction
    of accelerated blood clearance of PEGylated liposomes. J Control Release 115:243–250
Ishida T, Ichihara M, Wang X, Yamamoto K, Kimura J, Majima E, Kiwada H (2006c) Injection of
    PEGylated liposomes in rats elicits PEG-specific IgM, which is responsible for rapid elimina-
    tion of a second dose of PEGylated liposomes. J Control Release 112:15–25
Jain RK (1987) Transport of molecules across tumor vasculature. Cancer Metastasis Rev 6:
    559–593
Johnston SR, Gore ME (2001) Caelyx: phase II studies in ovarian cancer. Eur J Cancer 37:S8–14
Katsaros D, Oletti MV, Rigault de la Longrais IA, Ferrero A, Celano A, Fracchioli S, Donadio M,
    Passera R, Cattel L, Bumma C (2005) Clinical and pharmacokinetic phase II study of pegylated
    liposomal doxorubicin and vinorelbine in heavily pretreated recurrent ovarian carcinoma. Ann
    Oncol 16:300–306
Klibanov AL, Maruyama K, Torchilin VP, Huang L (1990) Amphipathic polyethyleneglycols
    effectively prolong the circulation time of liposomes. FEBS Lett 268:235–237
Koch G, Lok BD, van Oudenaren A, Benner R (1982) The capacity and mechanism of bone marrow
    antibody formation by thymus-independent antigens. J Immunol 128:1497–1501
Koukourakis MI, Koukouraki S, Giatromanolaki A, Archimandritis SC, Skarlatos J, Beroukas K,
    Bizakis JG, Retalis G, Karkavitsas N, Helidonis ES (1999) Liposomal doxorubicin and con-
    ventionally fractionated radiotherapy in the treatment of locally advanced non-small-cell lung
    cancer and head and neck cancer. J Clin Oncol 17:3512–3521
Krown SE, Northfelt DW, Osoba D, Stewart JS (2004) Use of liposomal anthracyclines in Kaposi's
    sarcoma. Semin Oncol 31:36–52
Laing RB, Milne LJ, Leen CL, Malcolm GP, Steers AJ (1994) Anaphylactic reactions to liposomal
    amphotericin. Lancet 344:682
Laverman P, Boerman OC, Oyen WJ, Dams ET, Storm G, Corstens FH (1999) Liposomes for
    scintigraphic detection of infection and inflammation. Adv Drug Deliv Rev 37:225–235
Laverman P, Carstens MG, Boerman OC, Dams ET, Oyen WJ, van Rooijen N, Corstens FH, Storm
    G (2001) Factors affecting the accelerated blood clearance of polyethylene glycol-liposomes
    upon repeated injection. J Pharmacol Exp Ther 298:607–612
Lee JH, Lee HB, Andrade JD (1995) Blood compatibility of polyethylene oxide surfaces. Prog
    Polymer Sci 20:1043–1079
Lotem M, Hubert A, Lyass O, Goldenhersh MA, Ingber A, Peretz T, Gabizon A (2000)
    Skin toxic effects of polyethylene glycol-coated liposomal doxorubicin. Arch Dermatol 136:
    1475–1480
Lum H, Malik AB (1994) Regulation of vascular endothelial barrier function. Am J Physiol
    267:L223–241
Lyass O, Uziely B, Ben-Yosef R, Tzemach D, Heshing NI, Lotem M, Brufman G, Gabizon
    A (2000) Correlation of toxicity with pharmacokinetics of pegylated liposomal doxorubicin
    (Doxil) in metastatic breast carcinoma. Cancer 89:1037–1047
Markman M, Kennedy A, Webster K, Peterson G, Kulp B, Belinson J (2000) Phase 2 trial of
    liposomal doxorubicin (40 mg/m(2)) in platinum/paclitaxel-refractory ovarian and fallopian
    tube cancers and primary carcinoma of the peritoneum. Gynecol Oncol 78:369–372
Martin F, Oliver AM, Kearney JF (2001) Marginal zone and B1 B cells unite in the early response
    against T-independent blood-borne particulate antigens. Immunity 14:617–629
Maruyama K, Yuda T, Okamoto A, Kojima S, Suginaka A, Iwatsuru M (1992) Prolonged circulation
    time in vivo of large unilamellar liposomes composed of distearoyl phosphatidylcholine and
    cholesterol containing amphipathic poly(ethylene glycol). Biochim Biophys Acta 1128:44–49
Mirchandani D, Hochster H, Hamilton A, Liebes L, Yee H, Curtin JP, Lee S, Sorich J, Dellenbaugh
    C and Muggia FM (2005) Phase I study of combined pegylated liposomal doxorubicin with
    protracted daily topotecan for ovarian cancer. Clin Cancer Res 11:5912–5919

Muggia FM, Hainsworth JD, Jeffers S, Miller P, Groshen S, Tan M, Roman L, Uziely B, Muderspach L, Garcia A, Burnett A, Greco FA, Morrow CP, Paradiso LJ, Liang LJ (1997) Phase II study of liposomal doxorubicin in refractory ovarian cancer: antitumor activity and toxicity modification by liposomal encapsulation. J Clin Oncol 15:987–993

Needham D, McIntosh TJ, Lasic DD (1992) Repulsive interactions and mechanical stability of polymer-grafted lipid membranes. Biochim Biophys Acta 1108:40–48

Northfelt DW, Martin FJ, Working P, Volberding PA, Russell J, Newman M, Amantea MA, Kaplan LD (1996) Doxorubicin encapsulated in liposomes containing surface-bound polyethylene glycol: pharmacokinetics, tumor localization, and safety in patients with AIDS-related Kaposi's sarcoma. J Clin Pharmacol 36:55–63

Northfelt DW, Dezube BJ, Thommes JA, Levine R, Von Roenn JH, Dosik GM, Rios A, Krown SE, DuMond C, Mamelok RD (1997) Efficacy of pegylated-liposomal doxorubicin in the treatment of AIDS-related Kaposi's sarcoma after failure of standard chemotherapy. J Clin Oncol 15: 653–659

O'Brien ME, Wigler N, Inbar M, Rosso R, Grischke E, Santoro A, Catane R, Kieback DG, Tomczak P, Ackland SP, Orlandi F, Mellars L, Alland L, Tendler C (2004) Reduced cardiotoxicity and comparable efficacy in a phase III trial of pegylated liposomal doxorubicin HCl (CAELYX/Doxil) versus conventional doxorubicin for first-line treatment of metastatic breast cancer. Ann Oncol 15:440–449

Oku N (1999) Delivery of contrast agents for positron emission tomography imaging by liposomes. Adv Drug Deliv Rev 37:53–61

Orditura M, Quaglia F, Morgillo F, Martinelli E, Lieto E, De Rosa G, Comunale D, Diadema MR, Ciardiello F, Catalano G, De Vita F (2004) Pegylated liposomal doxorubicin: pharmacologic and clinical evidence of potent antitumor activity with reduced anthracycline-induced cardiotoxicity (review). Oncol Rep 12:549–556

Overmoyer B, Silverman P, Holder LW, Tripathy D, Henderson IC (2005) Pegylated liposomal doxorubicin and cyclophosphamide as first-line therapy for patients with metastatic or recurrent breast cancer. Clin Breast Cancer 6:150–157

Papahadjopoulos D, Allen TM, Gabizon A, Mayhew E, Matthay K, Huang SK, Lee KD, Woodle MC, Lasic DD, Redemann C, Martin FJ (1991) Sterically stabilized liposomes: improvements in pharmacokinetics and antitumor therapeutic efficacy. Proc Natl Acad Sci USA 88: 11460–11464

Patel HM (1992) Serum opsonins and liposomes: their interaction and opsonophagocytosis. Crit Rev Ther Drug Carrier Syst 9:39–90

Phillips NC (1989) Kupffer cells and liver metastasis. Optimization and limitation of activation of tumoricidal activity. Cancer Metastasis Rev 8:231–252

Poh SB, Bai LY, Chen PM (2005) Pegylated liposomal doxorubicin-based combination chemotherapy as salvage treatment in patients with advanced hepatocellular carcinoma. Am J Clin Oncol 28:540–546

Safra T (2003) Cardiac safety of liposomal anthracyclines. Oncologist 8 Suppl 2:17–24

Safra T, Muggia F, Jeffers S, Tsao-Wei DD, Groshen S, Lyass O, Henderson R, Berry G, Gabizon A (2000) Pegylated liposomal doxorubicin (Doxil): reduced clinical cardiotoxicity in patients reaching or exceeding cumulative doses of 500 mg/m2. Ann Oncol 11:1029–1033

Sharpe M, Easthope SE, Keating GM, Lamb HM (2002) Polyethylene glycol-liposomal doxorubicin: a review of its use in the management of solid and haematological malignancies and AIDS-related Kaposi's sarcoma. Drugs 62:2089–2126

Szebeni J (2005a) Complement activation-related pseudoallergy caused by amphiphilic drug carriers: the role of lipoproteins. Curr Drug Deliv 2:443–449

Szebeni J (2005b) Complement activation-related pseudoallergy: a new class of drug-induced acute immune toxicity. Toxicology 216:106–121

Szebeni J, Fontana JL, Wassef NM, Mongan PD, Morse DS, Dobbins DE, Stahl GL, Bunger R, Alving CR (1999) Hemodynamic changes induced by liposomes and liposome-encapsulated hemoglobin in pigs: a model for pseudoallergic cardiopulmonary reactions to liposomes.

Role of complement and inhibition by soluble CR1 and anti-C5a antibody. Circulation 99: 2302–2309

Szebeni J, Baranyi L, Savay S, Bodo M, Morse DS, Basta M, Stahl GL, Bunger R, Alving CR (2000) Liposome-induced pulmonary hypertension: properties and mechanism of a complement-mediated pseudoallergic reaction. Am J Physiol Heart Circ Physiol 279:H1319–1328

Szebeni J, Baranyi L, Savay S, Milosevits J, Bunger R, Laverman P, Metselaar JM, Storm G, Chanan-Khan A, Liebes L, Muggia FM, Cohen R, Barenholz Y, Alving CR (2002) Role of complement activation in hypersensitivity reactions to doxil and hynic PEG liposomes: experimental and clinical studies. J Liposome Res 12:165–172

Torchilin VP, Omelyanenko VG, Papisov MI, Bogdanov AA Jr, Trubetskoy VS, Herron JN, Gentry CA (1994) Poly(ethylene glycol) on the liposome surface: on the mechanism of polymer-coated liposome longevity. Biochim Biophys Acta 1195:11–20

Tsai NM, Cheng TL, Roffler SR (2001) Sensitive measurement of polyethylene glycol-modified proteins. Biotechniques 30:396–402

Uziely B, Jeffers S, Isacson R, Kutsch K, Wei-Tsao D, Yehoshua Z, Libson E, Muggia FM, Gabizon A (1995) Liposomal doxorubicin: antitumor activity and unique toxicities during two complementary phase I studies. J Clin Oncol 13:1777–1785

Vaage J, Donovan D, Mayhew E, Abra R, Huang A (1993) Therapy of human ovarian carcinoma xenografts using doxorubicin encapsulated in sterically stabilized liposomes. Cancer 72: 3671–3675

Vaage J, Donovan D, Uster P, Working P (1997) Tumour uptake of doxorubicin in polyethylene glycol-coated liposomes and therapeutic effect against a xenografted human pancreatic carcinoma. Br J Cancer 75:482–486

van Rooijen N, van Nieuwmegen R (1980) Liposomes in immunology: multilamellar phosphatidylcholine liposomes as a simple, biodegradable and harmless adjuvant without any immunogenic activity of its own. Immunol Commun 9:243–256

Verhaar-Langereis M, Karakus A, van Eijkeren M, Voest E, Witteveen E (2006) Phase II study of the combination of pegylated liposomal doxorubicin and topotecan in platinum-resistant ovarian cancer. Int J Gynecol Cancer 16:65–70

Wang X, Ishida T, Kiwada H (2007) Anti-PEG IgM elicited by injection of liposomes is involved in the enhanced blood clearance of a subsequent dose of PEGylated liposomes. J Control Release 119:236–244

Waterhouse DN, Tardi PG, Mayer LD, Bally MB (2001) A comparison of liposomal formulations of doxorubicin with drug administered in free form: changing toxicity profiles. Drug Saf 24:903 920

Woodle MC (1998) Controlling liposome blood clearance by surface-grafted polymers. Adv Drug Deliv Rev 32:139–152

Woodle MC, Lasic DD (1992) Sterically stabilized liposomes. Biochim Biophys Acta 1113: 171–199

Wu NZ, Da D, Rudoll TL, Needham D, Whorton AR, Dewhirst MW (1993) Increased microvascular permeability contributes to preferential accumulation of Stealth liposomes in tumor tissue. Cancer Res 53:3765–3770

Yan X, Scherphof GL, Kamps JA (2005) Liposome opsonization. J Liposome Res 15:109–139

Yuan F, Dellian M, Fukumura D, Leunig M, Berk DA, Torchilin VP, Jain RK (1995) Vascular permeability in a human tumor xenograft: molecular size dependence and cutoff size. Cancer Res 55:3752–3756

Zandvoort A, Timens W (2002) The dual function of the splenic marginal zone: essential for initiation of anti-TI-2 responses but also vital in the general first-line defense against blood-borne antigens. Clin Exp Immunol 130:4–11

Chapter 7
Hydrogel Nanocomposites: Biomedical Applications, Biocompatibility, and Toxicity Analysis

Samantha A. Meenach, Kimberly W. Anderson, and J. Zach Hilt

Abstract Hydrogel nanocomposites are an important class of biomaterials that can be utilized in applications such as drug delivery, tissue engineering, and hyperthermia treatment. The incorporation of nanoparticles into a hydrogel matrix can provide unique properties including remote actuation and can also improve properties such as mechanical strength. Since hydrogel nanocomposites have been proposed as implantable biomaterials, it is important to analyze and understand the response of the body to these novel materials. This chapter covers the background, definitions, and potential applications of hydrogels and hydrogel nanocomposites. It also covers the various types of hydrogel nanocomposites as defined by the nanoparticulates embedded in the systems which include clay, metallic, magnetic, and semiconducting nanoparticles. The specific concerns of the biocompatibility analysis of hydrogel nanocomposites are discussed along with the specific biocompatibility results of the nanoparticulates incorporated into the hydrogel matrices as well as the biocompatibility of the hydrogels themselves. The limited data available on the biocompatibility of hydrogel nanocomposites is also presented. Overall, currently investigated hydrogel systems with known biocompatibility may have the potential to provide a "shielding" effect for the nanoparticulates in the hydrogel nanocomposites allowing them to be safer materials than the nanoparticulates alone.

Contents

J.Z. Hilt (✉)
Department of Chemical and Materials Engineering, University of Kentucky, Lexington, KY 40506-0046, USA
e-mail: hilt@engr.uky.edu

T.J. Webster (ed.), *Safety of Nanoparticles*, Nanostructure Science and Technology, DOI 10.1007/978-0-387-78608-7_7, © Springer Science+Business Media, LLC 2009

7.1 Introduction

Hydrogel nanocomposites are an important class of biomaterials that can be utilized in applications such as drug delivery, tissue engineering, and hyperthermia treatment. The incorporation of nanoparticles into a hydrogel matrix can provide unique properties including remote actuation and also improve properties such as mechanical strength. An advantage of using hydrogel nanocomposites in biomedical applications stems from their similarity to soft tissue as well as expected biocompatibility. Overall, the safety concern of hydrogel nanocomposites involves the biocompatibility of the system when implanted in the body. As such, it is important to test the effect of the nanocomposite on the body before it is utilized in any biomedical application. This chapter reviews the background concerning hydrogels and hydrogel nanocomposites along with the biocompatibility of the constituents that make up the composites. The limited data available on the biocompatibility of hydrogel nanocomposites is also presented.

7.2 Hydrogels

Hydrogels are an important class of polymeric materials that have been utilized in a wide variety of biomedical and pharmaceutical applications. Hydrogels are three-dimensional, hydrophilic, polymeric networks that can absorb up to thousands of times their dry weight in water or biological fluids (Hoffman 2002, Peppas et al. 2000). They consist of polymer chains with either physical or chemical crosslinks that prevent the dissolution of the hydrogel structure and instead result in swelling of the material upon interaction with aqueous solutions. Hydrogels are advantageous for many biomedical applications due to their resemblance of natural living tissue and inherent biocompatibility which can be partially attributed to their soft, flexible nature and high water content (Hoffman 2002). Hydrogel systems such as poly(hydroxyethyl methacrylate) (PHEMA), poly(N-isopropylacrylamide) (PNIPAAm), poly(vinyl alcohol) (PVA), and poly(ethylene glycol) (PEG) have been widely investigated for a wide variety of biomedical and pharmaceutical applications.

For many hydrogels, their swelling and deswelling behavior is dependent on external stimuli. Responsive hydrogels increase or decrease their swelling ratio as a result of changing the pH, temperature, ionic strength, or electromagnetic radiation of their environment (Peppas et al. 2000). For example, PNIPAAm is a temperature-responsive polymer that has been studied for many applications. It undergoes a reversible swelling transition around its lower critical solution temperature (LCST) of approximately 33°C in aqueous media. At temperatures below the LCST, PNIPAAm hydrogels are swollen whereas above this temperature they are collapsed. With the ability to control the behavior of these environmentally responsive hydrogels, they can be utilized in a wide variety of applications including drug delivery, tissue engineering, sensors, and actuators (Hoffman 2002; Peppas et al. 2000, 2006).

The biocompatibility of hydrogels has been thoroughly studied for a number of hydrogel systems and biological applications. As with any biomaterial, it has been necessary to make sure that the body responds favorably to the gel and that the body does not harm the material. For many hydrogel systems, these studies have been performed and have resulted in positive results. The expansion of hydrogel nanocomposite research in recent years has resulted in novel biomaterials with unique properties, but in addition, there is an increasing concern of their safety. Currently, there is limited literature concerning their biocompatibility. As with any material that hopes to go from the lab bench to the clinic, it is necessary that the safety and biocompatibility of hydrogel nanocomposite systems are studied in detail.

7.3 Hydrogel Nanocomposites

Despite the many advantages of using conventional crosslinked hydrogels, their applications are often limited due to their poor mechanical and limited response properties (Xiang et al. 2006). The random nature of the crosslinking reactions involved in hydrogel fabrication and the resulting morphological inhomogeneity can induce these limitations (Haraguchi et al. 2003). Recently, work has been done to improve hydrogel properties (e.g., mechanical strength) and to add unique properties (e.g., response to novel stimuli) through the fabrication of hydrogel nanocomposites (Frimpong and Hilt 2007). Hydrogel nanocomposites involve the incorporation of various nanoparticulate materials within a versatile hydrogel matrix which can provide easy, straightforward methods for enhancing the properties (e.g., improving the mechanical properties) of hydrogels. Although a number of fabrication techniques have been used to create such systems, in situ polymerization of particles into a monomer solution is a common way to create a hydrogel nanocomposite.

Nanocomposite hydrogels have been shown to modify and improve a variety of material properties, including magnetic and optical properties. For example, it is possible to tune a temperature-responsive system with electrochemical responses of a conducting polymer through the addition of electroactive, conducting particles in a hydrogel matrix (Kim et al. 2000). Thus far, a number of nanoparticulates have been utilized in nanocomposite hydrogel systems including metallic nanoparticles, carbon nanotubes (CNTs), clay, ceramics, magnetic nanoparticles, hydroxyapatite (HA), and semiconducting nanoparticles. Table 7.1 provides a brief summary of some of the nanocomposite systems that have been studied. In the following sections, hydrogel nanocomposites are classified by the nanoparticulate, and high interest examples are highlighted.

7.3.1 Clay Nanocomposites

One of the most widely studied classes of hydrogel nanocomposites involves the addition of nanoparticulate clay to the hydrogel system. Thermoresponsive PNIPAAm systems have been the most commonly used, however, systems involving poly(acrylic acid) (PAA), poly(methyl methacrylate) (PMMA), and poly(N, N-dimethylacrylamide) (PDMAA) have also been studied. One of the main advantages of the addition of clay to hydrogels is that the clay has been shown to act as a crosslinking agent, increasing the mechanical properties of the composites (Bandi et al. 2005; Santiago et al. 2007). The improvement in mechanical strength is primarily driven by good dispersion and/or the ability of the clay to exfoliate in the polymer (Bandi et al. 2005). The types of clay nanoparticulates incorporated into hydrogel nanocomposites include montmorillonite (MMT), bentonite, and other silicate clays. These systems have been widely characterized regarding changes in mechanical strength, swelling properties, drug release rates, and thermal transitions.

For many applications, an ideal PNIPAAm material is the one exhibiting structural homogeneity, high strength and toughness, and a high swelling ratio. For these composites to be used for applications such as rapid actuators, drug release devices, and artificial muscles, it is also important that they exhibit responsive properties such as rapid swelling and deswelling rates. Improvements in mechanical properties and gel transparency by changing the crosslinking clay amount in a PNIPAAm/inorganic clay hydrogel nanocomposite in comparison to conventionally crosslinked PNIPAAm hydrogels have been reported (Haraguchi et al. 2002; Haraguchi and Li, 2005). Liang et al. (2000) have shown a faster swelling transition and increased release rate with a PNIPAAm/clay nanocomposite hydrogel through successful dispersion of the clay throughout the nanocomposite. PNIPAAm/MMT nanocomposites have been shown to have increased mechanical strength in comparison to conventionally crosslinked hydrogels (Lee and Fu 2002; Lee and Ju 2004; Ma et al. 2007). For example, the shear modulus of this type of gel was shown to increase from 6.49×10^2 MPa to 11.42×10^2 MPa upon the addition of 15 wt% MMT versus pure PNIPAAm (Lee and Fu 2002).

Other clay–hydrogel nanocomposite systems have been studied including the combination of attapulgite fibrils, a hydrated magnesium aluminum silicate, with a poly(HEMA-poly(ethylene glycol) methacrylate) (PEGMA)-methyl acrylic acid(MAA)) hydrogel (Xiang et al. 2006) and the combination of clay with poly(N,N-dimethacrylate) (PDMAA) (Haraguchi et al. 2003). Each nanocomposite

Table 7.1 Summary of hydrogel nanocomposite systems, their applications, and key results

Type of nanocomposite	Nanoparticulate component	Hydrogel component	Application(s)	Key results	References
Clay	Attapulgite fibrils	Poly(HEMA-PEGMA-MAA)	Drug release, tissue engineering	Improved tensile strength, response rate, and swelling ratio	Xiang et al. (2006)
	Bentonite	PNaA	Drug delivery	Equilibrium swelling decreased, thermal stability increased	Santiago et al. (2007)
	Hectorite	PNIPAAm	Cell cultivation, tissue engineering	Cell adhesion and proliferation shown on composite	Haraguchi and Li (2005)
	Hectorite	PNIPAAm	Smart gels, enzyme carrier, drug delivery, tissue engineering	Swelling transition of PNIPAAm was better-controlled	Haraguchi et al. (2002)
	Laponite XLG	PDMAA	Drug delivery	Improved mechanical and swelling properties	Haraguchi et al. (2003)
	Laponite XLG	PNIPAAm	Drug delivery	Mechanical properties, transparencies, swelling, and deswelling behaviors changed by cross-linker contents	Haraguchi et al. (2006)
	Laponite XLG	PNIPAAm and CMCS	Controlled drug release, chemical valves	Improved swelling ratio, reponse rate, and mechanical properties	Ma et al. (2007)
	MMT	PMMA, PB, PS	Drug delivery	Improved mechanical properties	Zhao et al. (2003)
	MMT	PNIPAAm	Sensors, actuators, regulators, controlled release devices	Faster thermal transition, increased release rate	Liang et al. (2000)
	MMT	PNIPAAm	Sensors, catalyst systems, insulating materials	Structural integrity of material increased	Bandi et al. (2005)
	MMT	PNIPAAm	Drug delivery	Swelling ratio decreased, mechanical properties increased	Lee and Fu (2002)

Table 7.1 (continued)

Type of nanocomposite	Nanoparticulate component	Hydrogel component	Application(s)	Key results	References
	MMT	PNIPAAm, PAA	Drug delivery	Increased gel strength	Lee and Ju (2004)
	OMMT	PAA	Various	Increased thermal stability and equilibrium swelling ratio	Weian et al. (2005)
HA	HA	Bacterial cellulose	Biomimetic apatite formation, orthopedics	Synthesis of the composite mimics the natural biomineralization of bone	Hutchens et al. (2006)
	HA	PAA	Biomimetic bone, biomaterial replant applications	More homogeneous distribution of particles, improved tensile strength	Shen et al. (2007)
	HA	PVA	Tissue remineralization, cartilage repair and replacement	Increased mechanical strength	Sinha et al. (2007)
	HA	PVA	Artificial articular cartilage	Good thermal stability and improved homogeneity of HA	Fenglan et al. (2004)
	HA	PVA and collagen	Tissue engineering, bone repair	Exhibited high elasticity, composite provided porosity	Degirmenbasi et al. (2006)
	HA (micron-size)	HYAFF11	Bone tissue engineering	Improved mechanical properties, bioactivity, biocompatibility, and hemocompatibility	Giordano et al. (2006)
	HA, TCP (micron-sized)	HPMC	Injectable material for bone tissue engineering	Successful bone formation shown on hydrogel scaffold	Trojani et al. (2006)
	HA, α-TCP	HYAFF11	Bone substitute	Improved mechanical properties, bioactive	Sanginario et al. (2006)
Magnetic	Fe$_3$O$_4$ nanoparticles	Gelatin	Drug delivery	Smart on-off systems developed to deliver drugs	Liu et al. (2006)

Table 7.1 (continued)

Type of nanocomposite	Nanoparticulate component	Hydrogel component	Application(s)	Key results	References
	Fe_2O_3 (magnetite)	PNIPAAm	Externally responsive to magnetic field	Can target and separate beads with magnetic field	Xuli et al. (2000)
	Fe_3O_4 nanoparticles	PNIPAAm	Drug delivery, hyperthermia	Tunable swelling reponse	Frimpong et al. (2006)
Metallic	Ag nanoparticles	PAAm	Muscle-like actuators, biosensors, switchable electronics	Increased swelling ratio and lower electron transfer resistance	Saravanan et al. (2007)
	Au nanoparticles	PNIPAAm	Catalyst, biosensors, reponsive material, drug delivery	Specific control over Au nanoparticles, improved bulk properties	Wang et al. (2004)
	Au colloids and nanoshells	P(NIPAAm-co-Aam)	Microfluidics, valves	Gels collapse in reponse to light source	Sershen et al. (2005)
	Au–AuS nanoshells	PNIPAAm, Aam	Drug delivery	Light actuated drug delivery exhibited	Sershen et al. (2000)
	Au, Ag nanoparticles	PNIPAAm, GMA	Biomedical and electronic devices	Reversably changeable color shown due to nanoparticle size	Suzuki and Kawaguchi (2006)
	Cu and Ag ions	Hyaluronane	Artificial blood vessels	Cytocompatible, promote vessel growth	Barbucci et al. (2002)
Semiconducting	CdS	PAAm	Signal-triggered photoelectrochemical systems	Solvent-induced switchable photoelectrochemical system	Pardo-Yissar et al. (2002)
	CdS	PDMAA	Two-phase catalysis, drug delivery	Different physical states of hydrogel were achieved	Bekiari et al. (2004)
	CdTe, CdSe (QDs)	PEG	Biosensors, fluorimmunoassays, wound healing	QDs entrapped and luminescence characteristics were shown	Gattas-Asfura et al. (2003)

Table 7.1 (continued)

Type of nanocomposite	Nanoparticulate component	Hydrogel component	Application(s)	Key results	References
Other	Bacterial cellulose fibers	PVA	Cardiovascular soft tissue replacement	Mechanical properties shown similar to heart valve tissue	Millon and Wan (2005)
	CaCO$_3$	PAAm modified with PAA	Biomimetic material, biomineralization	Crystalline assembly of material for bone engineering	Grassmann and Lobmann (2004)
	Chitosan lactate	PVA	Wound dressing material	Improved biocompatibility, hemocompatibility, mechanical properties	De Queiroz et al. (2002)
	Hydrotalcite	poly(AA-co-NIPAAm)	Drug Delivery	Increased swelling ratio in saline solution but lower in alcohol solution	Le and Chen (2006)
	MWCNT	PVA	Microswitches, artificial muscle, prosthetic devices	Electrically actuated hydrogel	Shi et al. (2005)
	PAAm, BSA, silica nanoparticles	PAAm	Auto-focusing intra-occular lens	PAAm with silica most promising for intra-occular lens applications	Prokop et al. (2002)
	Polymer colloids (PPy)	PAA, glycerol	Sensors, membranes, biomaterials, actuators	Hydrogels shown to be electroactive	Kim et al. (2000)
	Silica nanoparticles	PHEMA	Tissue engineering	Higher swelling ratio, bioactive	Costatini et al. (2006)
	Silica nanoparticles	PHEMA	Bone tissue engineering	Improved supportment of osteoclasts and cytocompatibility	Schiraldi et al. (2004)
	TiO$_2$ nanoparticles	PAAm	Degradation of organic pollutants	Showed good photocatalytic degradability with reproducibility	Tang et al. (2007)
	TiO$_2$ nanoparticles	PHEMA (sol-gel)	Bone tissue engineering	Bioactive: HA formation on surface	Prashantha et al. (2006)

system showed improvements in mechanical and swelling properties such as increased strength and faster response rates and swelling ratios versus conventional hydrogels. Clay has also been used with PAA and PMMA, and these systems showed improved properties such as higher thermal stability and uniform dispersion (Zhao et al. 2003; Weian et al. 2005). Overall, the improved hydrogel properties that result from the introduction of nanosized-clay will allow for improved material performance and for use in new applications.

7.3.2 Hydroxyapatite Nanocomposites

Hydroxyapatite (HA) nanocomposites can potentially be used in a wide variety of applications including bone tissue engineering as well as cartilage repair and replacement, and injectable materials for tissue engineering. HA is a calcium phosphate-derived mineral commonly used as a biomaterial (Hutchens et al. 2006). HA acts as a biomimetic material that induces bone growth on its surface owing to its bioactivity. Generally, hydrogels have not been used in bone tissue engineering applications due to their low mechanical strength and lack of bone growth stimulation. The incorporation of HA in a hydrogel nanocomposite can potentially overcome these limitations. HA has been combined with many hydrogel systems including those comprised bacterial cellulose, PAA, PVA, and collagen (Hutchens et al. 2006, Shen et al. 2007, Sinha et al. 2007, Fenglan et al. 2004, Degirmenbasi et al. 2006). The most important properties that should be exhibited in HA hydrogel nanocomposites include improved biocompatibility, bioactivity, mechanical properties, thermal stability, and elasticity.

Incorporating nanosized HA nanoparticulates into PVA hydrogels has shown to increase the mechanical strength of the gel, provide good thermal stability and homogeneity, increase elasticity, and provide porosity for different systems (Sinha et al. 2007, Fenglan et al. 2004, Degirmenbasi et al. 2006). HA-PVA hydrogels were synthesized by Sinha et al. (2007) which formed a macroporous nanocomposite block of the HA nanoparticles. Fenglan et al. (2004) investigated a PVA-HA nanocomposite which showed good homogeneity and thermal stability and can be used as a high-performance material for artificial articular cartilage due to the bioactivity the HA provides.

Studies on HYAFF®11, a biocompatible and biodegradable benzyl ester of hyaluronic acid, have also showed good results in improving the properties of hydrogels. Giordano et al. (2006) incorporated HYAFF®11 with micron-sized HA which showed improved mechanical properties, bioactivity, and biocompatibility. In specific, this system was shown to have compressive strengths similar to spongy bone where the maximum compressive strength was 12 MPa for the dry system and 4 MPa for the swollen system. This work was extended to nanosized HA incorporated in HYAFF®11 by Sanginario et al. (2006) which showed results similar to the micron-sized HA in which the HA powders were shown to improve the mechanical properties of HYAFF®11. These gels can be used to promote the bioactivity of the material to facilitate bone growth as needed.

Shen et al. developed a template-driven method for the synthesis of PAA and HA nanocomposites. This method demonstrated homogeneous distribution of the particles throughout the material which improved the strength of the gel (Shen et al. 2007). A micron-sized HA system that has been studied and can potentially be extended to a nanosized HA system includes a new synthetic injectable hydrogel fabricated by Trojani et al. (2006). This nanocomposite system was made from Si–hydroxypropylmethylcellulose (HPMC) with HA/tricalcium phosphate particles and showed in vivo biocompatibility with successful new bone formation efficacy when loaded with undifferentiated bone marrow stromal cells (BMSCs). Bone tissue engineering is a very important and widely studied field, and thus, the utilization of hydrogels in this application is ongoing.

7.3.3 Magnetic Nanocomposites

The incorporation of magnetic nanoparticles such as iron oxide nanoparticles into hydrogels can create tunable nanocomposites that can be remote controlled by a magnetic field and have a wide variety of potential applications. These applications range from controlled drug release applications to hyperthermia treatments in cancer patients. One of the most advantageous aspects in using magnetic nanocomposites is their potential to have remote-controlled responsive properties upon exposure to external magnetic fields. For example, Liu et al. developed a gelatin/Fe_3O_4 hydrogel nanocomposite with uniform distribution of the magnetic nanoparticles throughout the gelatin matrix. This system exhibited a drug release profile controllable by switching between the on and off modes of a given DC magnetic field (Liu et al. 2006). When the magnetic field is on, the magnetic nanoparticles aggregated causing a decrease in porosity of the gel which in turn reduced the swelling rate and drug release rate of the hydrogel. Figure 7.1 shows the mechanism of the "on" configuration of the ferrogels due to aggregation of the iron oxide nanoparticles along with the swelling rate profile of a sample system.

PNIPAAm/Fe_2O_3 nanocomposite hydrogels which were studied by Xuli et al. (2000) also exhibited an external response to a magnetic field. Small pieces of the gel were cut and characterized for temperature and magnetic field sensitivities and it was shown that the gel beads could be targeted and separated by the magnetic field. Work has also been done in our laboratories on magnetically responsive hydrogel networks based on magnetic nanoparticles (Fe_3O_4) and PNIPAAm. Frimpong et al. (2006) have shown a tunable swelling response of this composite based on the type and amount of crosslinker used in the system. The presence of iron oxide particles in these hydrogel nanocomposites did not significantly alter the temperature sensitivity of these systems as is shown in Fig. 7.2. In recent work by Satarkar et al., we have demonstrated remote-controlled heating and drug release using hydrogel nanocomposites, as is shown in Fig. 7.3. Overall, the addition of magnetic nanoparticles into hydrogel matrices has the potential to create remote-controlled biomaterials for implant applications, such as in hyperthermia therapy and controlled drug release.

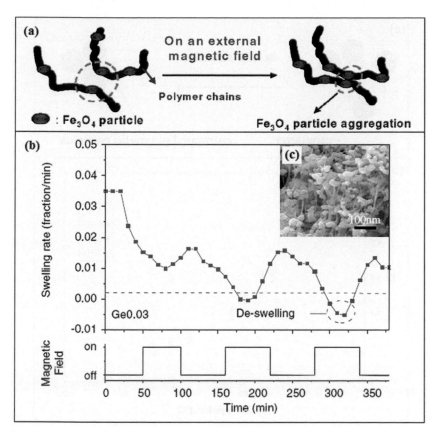

Fig. 7.1 (**a**) Mechanism of a "close" configuration of ferrogels due to the aggregation of Fe_3O_4 nanoparticles under an "on" magnetic field causes the porosity of the ferrogels to decrease, (**b**) swelling rate of ferrogels is dependent on switching magnetic fields, and (**c**) SEM observation of Fe_3O_4 nanoparticles distributed in gelatin hydrogels (Liu et al. 2006)

7.3.4 Metallic Nanocomposites

The synthesis of hydrogel nanocomposites incorporating metal nanoparticles and ions has widely increased owing to their unique optical, electrical, and catalytic properties (Saravanan et al. 2007). The metallic particles can provide composites with the ability to be actuated by light as well as decreasing the electrical resistance of the material in some applications. These materials can then be used in applications such as muscle-like actuators, biosensors, drug delivery, and switchable electronics. Metallic nanoparticles such as gold and silver have been incorporated in hydrogel systems such as those of PAAm and PNIPAAm (Saravanan et al. 2007; Wang et al. 2004). In certain cases, these nanocomposites have the ability to be remotely controlled via light actuation. For example, a system of Au colloids and nanoshells has been developed in which the hydrogel system collapsed in response

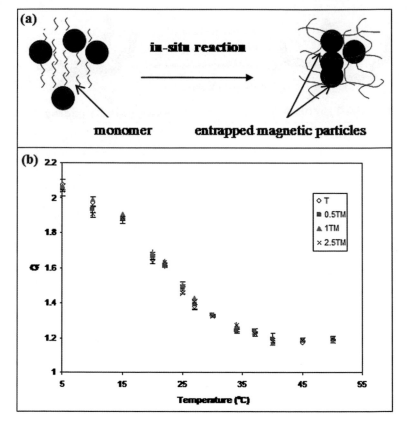

Fig. 7.2 (a) Schematic of in-situ polymerization of magnetic nanoparticles with a monomer solution to form magnetic-hydrogel nanocomposites. (b) Magnetic nanoparticle concentration effects on equilibrium volume swelling ratio for TEGDMA systems ($N = 3$, $\pm$SD) (Frimpong et al. 2006)

to a specific wavelength of light (Sershen et al. 2005). Figure 7.4 illustrates the swelling response of the hydrogel nanocomposite along with its application into a microfluidic device. A similar system also developed by Sershen et al. (2000) comprised P(NIPAAm-co-AAm) and Au-AuS nanoshells showed photothermally modulated drug delivery. The Au–AuS nanoparticles strongly absorbed near-infrared light, and when they were incorporated into the hydrogel NC and exposed to wavelengths between 800 and 1200 nm, the light absorbed by the nanoparticles was converted to heat. This increased the temperature above the LSCT of the PNIPAAm system and consequently a drug was released from the nanocomposite matrix.

Nanocomposite hydrogels containing metal nanoparticles have also been used in biosensor applications and to promote proangiogenic activity in blood vessel formation. Suzuki and Kawaguchi developed a PNIPAAm and glycidyl methacrylate copolymerized hydrogel system infused with Au and Ag nanoparticles which

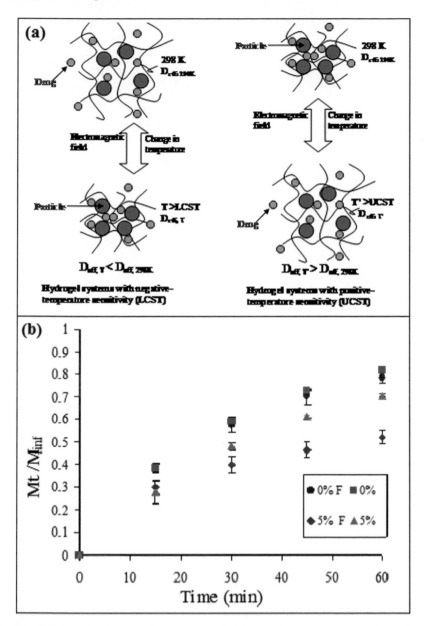

Fig. 7.3 (**a**) Schematic of proposed remote-controlled drug delivery system for negative and positive temperature sensitive systems. (**b**) Controlled drug release from nanocomposites in an electromagnetic field. % represents particle loading by weight in NIPAAm-PEG400DMA nanocomposite. F represents the application of a magnetic field. ($N = 3$, $\pm$SD) (Satarkar et al. 2007)

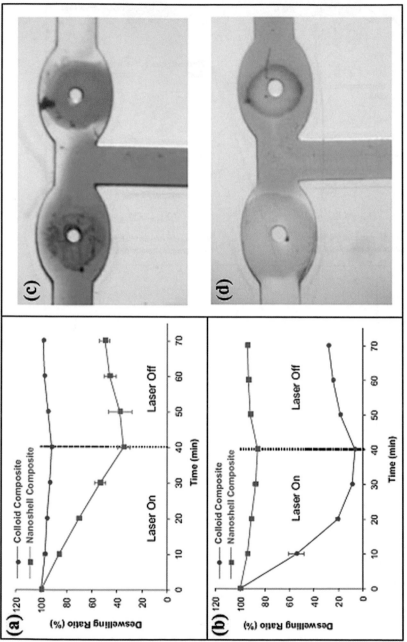

Fig. 7.4 The collapse and reswelling of the gold-colloid composite hydrogels (green squares) during and after 40 min of irradiation at (**a**) 832 nm (2.7 W cm^{-2}) or (**b**) 532 nm (1.6 W cm^{-2}). Two valves formed at a T-junction in a microfluidics device, one made of a gold-colloid nanocomposite hydrogel and the other a gold-nanoshell nanocomposite hydrogel. The channels are 100 μm wide. (**c**) When the entire device was illuminated with near infrared light (832 nm, 2.7 W cm^{-2}) the nanoshell valve opened while the gold colloid valve remained closed. However, when the device was illuminated with green light (532 nm 1.6 W cm^{-2}), the opposite reponse was observed. In both cases, the valves opened within 5 sec (Sershen et al. 2005)

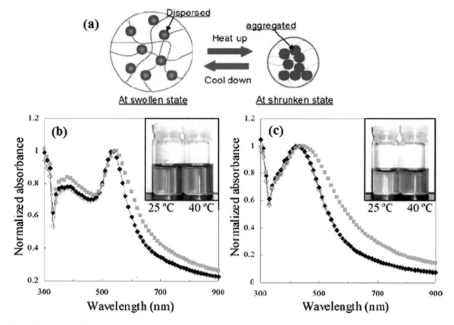

Fig. 7.5 (**a**) Schematic of thermosensitive color change by interparticulate interactions using structured nanoparticles, (**b**) UV–vis spectra of Hydrogel-NH$_2$-Au/Au2, and (**c**) Hydrogel-NH$_2$-Au/Ag/Au microgels measured at 25°C (black diamonds), 40°C (gray squares), and 25°C after 10 heating-cooling cycles (white diamonds) (Shi et al. 2005)

exhibited reversible multiple colors due to interparticulate interactions of surface plasmon resonance (SPR) using various nanoparticles (Suzuki and Kawaguchi 2006). The color change reversibility of the hydrogel NC was accomplished by adjusting the size of the nanoparticles. Figure 7.5 demonstrates the swelling behavior of this hydrogel composite along with the visible results due to the heating of the systems. In another hydrogel system, copper (Cu^{2+}) and silver (Ag$^+$) ions were integrated in a hyaluronane-based hydrogel nanocomposite by Barbucci et al. (2002). These hydrogel NCs showed proangiogenic activity through the stimulation of the growth of new blood vessels. Consequently, metallic hydrogel nanocomposites can be utilized in a wide variety of applications due to the improved tunability of drug release and proangiogenic activity within the systems.

7.3.5 Semiconducting Nanocomposites

A limited amount of work has been done on semiconducting hydrogel nanocomposites in which quantum dots (QDs) have been incorporated into hydrogel matrices. The strong interest in quantum dots stems from their unique luminescent properties, which leads to potential applications in a wide variety of fields including optoelectronics, biological immunoassays, and biosensors (Gattas-Asfura et al. 2003). The size of the semiconductor nanoparticulate materials is what provides the ability to customize luminescence for a particular application. The optoelectronic properties are due to quantum-confinement effects that occur when a semiconductor particle size falls within the bulk exciton Bohr radius (Gattas-Asfura et al. 2003). QDs have several important advantages including size-selection, narrow emission spectra, resistance to photobleaching, improved brightness, and broad excitation spectra. This can allow for semiconducting nanocomposites to be used in long-term sensing applications.

Pardo-Yissar et al. (2002) demonstrated solvent-induced switchable photoelectrochemical functions of hydrogel composites of PAAm and CdS that can be used for signal-triggered systems. Bekiari et al. fabricated a PDMMA/CdS hydrogel nanocomposite which could potentially be used in two-phase catalysis or drug delivery applications. Another hydrogel composite was synthesized by Gattas-Asfura et al. through the immobilization of CdTe and CdSe quantum dots in PEG via physical entrapment which endowed the hydrogel with luminescent properties (Gattas-Asfura et al. 2003). A CdS/PDMAA system has also been shown to be useful in drug delivery applications (Bekiari et al. 2004). Further investigation of systems involving quantum dots must be performed in order to continue discovering new and better uses for these versatile nanoparticulate materials.

7.3.6 Miscellaneous Nanocomposites

Along with the previously discussed hydrogel nanocomposites, there are a number of other nanoparticulates that have been used in hydrogel systems such as titanium

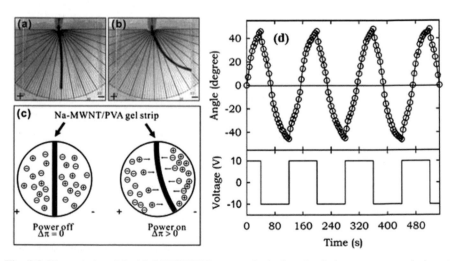

Fig. 7.6 Photographs of the Na-MWNT/PVA composite hydrogel strip in an aqueous solution at 10 V/cm DC electric field at (**a**) 0 s and (**b**) 30 s. In (**c**) the schematic illustration of the bending mechanism of the hydrogel strip due to the change of the osmosic pressure under a DC electric field is shown (Barbucci et al. 2002)

oxide, CNTs, chitosan, polymer colloids, and silica (De Queiroz et al. 2002; Millon and Wan 2005; Tang et al. 2007; Prokop et al. 2002; Kim et al. 2000). Depending on the type of nanoparticle used, they have been shown to change the properties of hydrogels for applications in areas such as bone tissue engineering, drug delivery, and wound dressing.

For bone tissue engineering applications, titanium oxide, silica, and $CaCO_3$ were utilized. PAAm gels were modified with $CaCO_3$ resulting in a nanocomposite hydrogel that incorporates an organic matrix into a crystalline assembly that is similar to the natural bone biomineralization processes (Grassmann and Lobmann 2004). PHEMA/SiO_2 nanocomposites have been developed which exhibit bioactivity and promote osteoblast growth on their surface (Schiraldi et al. 2004; Costatini et al. 2006). The silica nanoparticles also improved the mechanical strength of the material with improved organization of the polymer network due to the hydrogen bonding of the polymeric chains with the hydrophilic particles. Prashanth et al. also showed that a nanocomposite hydrogel consisting of PHEMA and TiO_2 nanoparticles fabricated using an in situ sol-gel process has in vitro bioactive properties through the formation of apatite on the surface of the material (Prashantha et al. 2006).

Carbon nanotubes can also be utilized in hydrogel nanocomposites. CNTs have high tensile strength, are ultra light weight, and have excellent chemical and thermal stability (Smart et al. 2006). They can be used as nano-fillers in hydrogel nanocomposites to dramatically improve mechanical properties and create highly anisotropic composites. CNTs can be used to create electronically conductive polymers and tissue engineering constructs with the capacity to provide controlled electrical stimulation. Shi et al. developed a novel type of actuator based on multi-walled carbon nanotubes (MWCNT) and PVA which was externally actuated using a DC electrical field (Shi et al. 2005). Figure 7.6 illustrates the bending mechanism of the hydrogel strip due to the change of the osmosic pressure under a DC electric field. This material can be used as a microswitch, artificial muscle, or prosthetic device.

7.4 Biocompatibility and Toxicity Analysis

Although a handful of hydrogel nanocomposite systems have been studied for their biocompatibility and bioactivity, this characterization has been left mostly unstudied, and thus, their safety in biomedical applications is unknown. Fortunately, a large body of work has been done in characterizing the hydrogels themselves as well as many of the nanoparticulate materials that are being used in these composites. The remaining portion of this chapter will focus on the biocompatibility results of hydrogels alone, the nanoparticles and nanoparticulates, and some hydrogel nanocomposites.

The issues surrounding the safety of hydrogel nanocomposites ultimately lie in the interest of using them in applications such as tissue engineering and drug delivery where the hydrogel matrices will be implanted in the body. Although there may be some concern in the handling and fabrication of these biomaterials, our concern focuses mainly on the safety of these composites regarding their biocompatibility. Classical chemical compounds are routinely tested for toxicity prior to release to the public whereas no such procedures exist for nanomaterials. Therefore, pre-screening nanomaterials are advantageous in determining detrimental effects on humans (Brunner et al. 2006). Despite the fact that a large number of hydrogel nanocomposites have been characterized for uses in a wide variety of biomedical applications, the literature available on their inherent biocompatibility is minimal.

Biocompatibility is defined by Williams as, "the ability of a material to perform with an appropriate host response in a specific application" (Williams 1986). The premise behind this is not only that a biomaterial initiates the desired response in the body but also that the body does not adversely affect the material itself. The desired host response to a biomaterial may include normal healing around the implant, resistance to blood clotting, or successful cell growth for tissue engineering. For some hydrogel nanocomposite implant applications, it is important that the gel be relatively inert, non-toxic, and not harmful to surrounding tissues.

It is necessary to perform a number of different tests on biomaterials in order to assess their biological tolerance and biocompatibility. This testing may involve in vivo characterization using animal models such as rats or in vitro testing conducted with cell culture lines. Owing to the fact that in vivo animal testing is expensive and involves regulatory concerns, in vitro biocompatibility testing is most often used to characterize newly developed biomaterials. Many in vitro biocompatibility tests have been developed to simulate and predict biological reactions to biomaterials when they are placed near tissues in the body (Hanks et al. 1996). These initial screening tests may include, but are not limited to, cytocompatibility or hemocompatibility analysis. Cytocompatibility analysis involves cytotoxicity tests which entail morphological evaluations on cell lines to determine if cells have died or undergo other negative changes because of interactions with the biomaterial (Pizzoferrato et al. 1994). Hemocompatibility evaluates the blood clotting cascade on biomaterials or devices that are in contact with blood.

When evaluating cytocompatibility, cell culture models can be used to screen biomaterials for acute toxicity and to assess any potentially harmful substances that may leach from the material. The cell lines used may be primary cells obtained directly from host tissues or continuous cell lines that are readily available from various companies such as the American Type Culture Collection (ATCC) (Pizzoferrato et al. 1994). Continuous cell lines such as fibroblasts or embryo cells are the most commonly used because they are usually inexpensive and easy to obtain. For cytocompatibility examination, the tests to be performed can be divided into three categories: (1) direct contact testing where the cells and material are in intimate contact, (2) indirect testing where the cells and materials are in the same solution but not touching, and (3) elution or extract testing which involves characterizing the effect of leachable substances in media that was in contact with the biomaterials from the material on the cells.

Hemocompatibility examination is also important in determining the biocompatibility of blood-contacting materials. From the International Organization of Standards perspective, there are five categories that can be explored for hemocompatibility evaluation including thrombosis, coagulation, platelet interactions, hematology, and immunology (Ratner et al. 2004). Despite this, there are no widely recognized standards for determining the blood compatibility of a material. Of the hemocompatibility testing performed, however, the evaluation methods most commonly utilized include measuring platelet adhesion and/or activation and measuring the mass of thrombus formation on the material. All of these evaluation methods are important because upon biomaterial interactions with injured tissue, platelets become activated, induce fibrin to crosslink, and form a platelet-fibrin thrombus which shows a lack of hemocompatibility of the material (Amarnath et al. 2006).

In regards to the biocompatibility of hydrogel nanocomposites, a limited amount of cytocompatibility or hemocompatibility studies have been performed. However, a wide variety of literature is available regarding the biocompatibility of the hydrogel and nanoparticulate components that make up a hydrogel nanocomposite. One of the advantages of hydrogel nanocomposites is that hydrogels can potentially provide increased biocompatibility over exposed nanoparticulates by encapsulating the particulate matter in the composite matrix and providing a barrier to sensitive tissues. Despite this, as more and more hydrogel nanocomposites are fabricated and characterized for various applications, it is important that continuous biocompatibility issues and safety of these materials be examined.

7.4.1 Biocompatibility and Toxicity Analysis of Hydrogels

Hydrogels comprised polymers such as PHEMA, PEG, and PNIPAAm have been readily investigated for biocompatibility and are currently used in biomedical applications such as contact lenses and tissue engineering. Other hydrogel systems including those of poly(vinyl alcohol) (PVA), poly(ethylene oxide) (PEO), poly(methyl methacrylate) (PMMA), and a combination of other polymers have also been studied for both cytotoxicity and hemocompatibility. Based on these studies, it is widely accepted that hydrogels are biocompatible and can be successfully used in biomedical applications. The biocompatibility of hydrogels stems from their similarity to soft tissue due to their hydrophilic nature and high water content.

For hydrogel systems, the cytotoxicity of the following could be examined: (1) the hydrogel constituents including the monomers, crosslinking agents, and initiators, (2) the substances leached from the crosslinked hydrogels (unreacted residues and byproducts), and (3) degradation products from the biodegradable hydrogels (Shin et al. 2003). Most toxicity issues surrounding hydrogels are associated with unreacted monomers and initiators that can leach out during application (Peppas et al. 2000). Some measures taken to overcome this issue include modifying the polymerization conditions to ensure higher conversion and extensive washing of the resulting hydrogel. Although biocompatibility testing of each system must be performed, the testing of a specific hydrogel system is sufficient due to the combination of the effect of the various constituents. A brief summary of the biocompatibility of hydrogels follows.

PHEMA is a highly biocompatible hydrogel that has been used extensively in specific biomedical applications such as contact lenses, vascular prostheses, drug delivery, and soft tissue replacement (Costatini et al. 2006). It has been combined with a number of other polymers to extend these applications and improve the characteristics of the materials. A hydrogel system of PEG and PHEMA was investigated for its use in blood-contacting medical devices and it was shown that the incorporation of PEG allowed for better blood biocompatibility (Bajpai 2007). This can be attributed to the fact that PEG is a hydrophilic polymer which has the ability to prevent protein adsorption which in turn minimizes blood clot formation.

PEG is an important biomaterial, with excellent biocompatibility properties. Its versatility has established the material as a desirable medium in many different biomedical applications (Gattas-Asfura et al. 2003). Selecting PEG as one of the components of a hydrogel is wise in that: (1) it is known to prevent the immunoreaction of the body to the biomaterial (stealth properties) (Yu et al. 1991), (2) the attachment of PEG to a biomaterial surface reduces protein adsorption and platelet adhesion (Bajpai 2007), and (3) it is a water-soluble polymer with high non-toxic and non-immunogenic properties (Bajpai and Shrivastava 2002).

Hydrogel systems of temperature-sensitive PNIPAAm have also been shown to be biocompatible. A PNIPAAm/poly(vinylcaprolactam) hydrogel was tested for in vitro cytotoxicity using two human carcinoma cell lines and it was found that the system was well tolerated by the gels (Vihola et al. 2005). PNIPAAm/hydroxypropyl cellulose nanoparticles were fabricated, and when tested in a mouse implantation model, it was found that they triggered minimal inflammatory cell accumulation and fibrotic capsule formation around the implantation site (Weng et al. 2004). One of the most important reasons for testing the biocompatibility of PNIPAAm hydrogel systems is that it is known that NIPAAm monomer may be carcinogenic or teratogenic (Weng et al. 2004).

A number of other tests including direct and indirect contact methods have been completed on other hydrogel systems. Dextran-based methacrylate hydrogels have shown good biocompatibility in vitro (De Groot et al. 2001), PAAm had no cytotoxic effects on NIH 3T3 murine fibroblasts, and multiple 2-methacryloyloxyethyl phosphorylcholine (MPC)-based systems have been found to be biocompatible when exposed to multiple cell lines (Kimura et al. 2006, Trojani et al. 2006, Nakabayashi and Williams 2003). Sugiyami et al. presented the biocompatibility of polyethylene grafted (g-PE) films modified with multiple polymer segments by measuring the adsorption of serum proteins to the surfaces and the rate of enzymatic reaction of thrombin with a synthetic substrate. It was found that PMPC-g-PE suppressed the amount of proteins adherent to the surface more than any other system because the PMPC segments have the largest amount of free water around the main chain (Sugiyami et al. 2000).

Overall, the hydrogel component of hydrogel nanocomposites typically poses the least risk when examining the composites for biocompatibility. Many hydrogels have been characterized for cytotoxicity and hemocompatibility as well as characterized using in vivo models. The ultimate success in their biocompatibility has led to their current use in applications such as tissue engineering and contact lenses.

7.4.2 Biocompatibility and Toxicity Analysis of Nanoparticulates

Nanoparticles may be considered the most "important" constituent in hydrogel nanocomposites in that they ultimately provide an improvement in physical applications and subsequent expansion of hydrogel applications. They are also of the greatest concern in evaluating the safety of the hydrogel systems. A wide variety of literature is available concerning the biocompatibility of nanoparticles. The biocompatibility of the nanoparticulate material depends on a number of characteristics specific to the material including size, chemical composition, shape, and surface interactions. Unfortunately, despite the information available on nanoparticle biocompatibility, much of it can be contradictory and misleading which is yet another reason why the evaluation of each new hydrogel nanocomposite for biocompatibility is so important. Specific testing is important because there are a number of ways nanoparticles can harm an organism. Nanoparticles could be composed of toxic materials that are harmful to cells. They could also cause harm by either sticking to the cells or due to cellular uptake of the particles. Finally, the shape of the nanoparticles could damage the cells (Kirchner et al. 2005).

Literature regarding biocompatibility is available on the most common nanoparticulate materials utilized in hydrogel nanocomposites including magnetic nanoparticles, semiconducting and metallic nanoparticles, and CNTs. For example, the use of gold nanoparticles has attracted a great deal of interest due to their interesting optical properties and potential applications in bioanalytical applications and PEG-coated gold nanoparticles have been found to be biocompatible due to the functionalized PEG monolayer on their surface (Tshikhudo et al. 2004).

Semiconducting nanoparticles (quantum dots or QDs) have been some of the more extensively studied nanoparticulate materials. Overall, there have been mixed reports on the stability and subsequent biocompatibility of quantum dots when used for long-term applications. In order to overcome quantum dot cytotoxicity, the release of Cd atoms into the surrounding medium must be inhibited and the biochemical mechanism of cytotoxicity must be determined (Tsay and Michalet 2005). Kirchner et al. studied CdSe and CdSe/ZnS quantum dots and showed that their cytotoxic effect was dependent on the surface chemistry of the nanoparticles and their stability towards aggregation (Kirchner et al. 2005). Coating QDs with ZnS increased the nanoparticle concentration by an order of 10 to where there were not toxic effects on the cells compared to uncoated QDs. It is possible that by providing a "shielding" effect to QDs with a hydrogel matrix that these systems could be rendered biocompatible.

Carbon nanotubes have also been studied for biocompatibility. It has been shown that rope-like agglomerates induce a more pronounced cytotoxic effect on cells than suspended CNT-bundles indicating a dependency on CNT aggregation (Wick et al.

2007). Also, Smart et al. (2006) presented a detailed review on the biocompatibility of CNTs including a wide array of contradictory materials. Despite this they were able to conclude that unrefined CNTs possess some degree of toxicity both in vivo and in vitro, that pristine CNTs have been shown to cause minimal cytotoxicity, and that chemically functionalized CNTs have not demonstrated any cytotoxicity. Ultimately, they warn that the scientific community should remain cautiously enthused by the potential biomedical applications of CNT-based materials until further more specific studies are conducted.

Iron oxide nanoparticles have been utilized in biomedical applications such as magnetic resonance imaging indicating their ability to be biocompatible. This biocompatibility is dependent on the coating surrounding the iron oxide nanoparticle. Iron oxide nanoparticles vary in their sizes and types of coatings which effects their circulation time in the blood, distribution in the body, and extent of uptake into cells (Muller et al. 2007). A number of coatings have been utilized and shown to render such magnetic particles biocompatible. These include polyoxyethylene (10) oleyl ether (Kim et al. 2001), PEG (Gupta and Curtis 2004), Frumoxtran-10 (Muller et al. 2007), hydroxide groups incorporated with $(CH_3)_4N^+$ (Cheng et al. 2005), and pullulan (Gupta and Gupta 2005). It is possible that by incorporating these materials into a hydrogel matrix that it may provide the necessary "coating" for the nanoparticles which can increase their biocompatibility.

A number of other nanoparticles used in hydrogel nanocomposites have been studied for biocompatibility including silica, HA, and titanium oxide. Overall, they have shown to be biocompatible depending on the experimental conditions. As such, for all nanoparticles investigated, it cannot be assumed that they are biocompatible until testing is done in the condition the nanoparticles will be used.

7.4.3 Biocompatibility and Toxicity Analysis of Hydrogel Nanocomposites

A limited number of hydrogel nanocomposites have been studied for safety in biomedical applications. One of the most significant studies involves a clay/hydrogel composite that can be used for cell cultivation (Haraguchi et al. 2006). In these studies, cell cultivation on the surface of a PNIPAAm/clay crosslinked network was shown to be successful for three lines of cells including HepG2 human hepatoma cells, human dermal fibroblasts, and human umbilical vein endothelial cells. A HYAFF®11 and HA nanocomposite showed no cytotoxic effects and no inhibition of osteoblast (bone cell) proliferation (Giordano et al. 2006). A PHEMA/silica composite also showed improved adhesion and proliferation of osteoblasts on the material with the increase of nanomeric filler content (Schiraldi et al. 2004). A self-hardening Si-hydroxypropymethylcellulose/HA/tricalcium phosphate hydrogel loaded with BMSCs was found to lack a fibrous capsule upon implantation indicating favorable biocompatibility (Trojani et al. 2006). Similarly, a PVA/chitosan composite was investigated for the use as a wound-dressing material and showed good results in in vivo experiments with rats (De Queiroz et al. 2002).

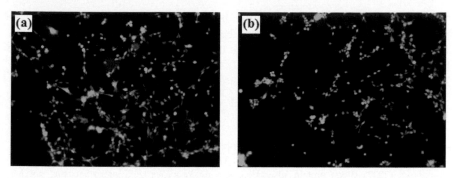

Fig. 7.7 Fluorescence microscopy images of live cells on a hydrogel nanocomposite comprised of 90 mol % PNIPAAm and 10 mol % TEGDMA at a seeding density of 100 cells cm^{-2} for 7 days (**a**) with 1 wt% Fe$_3$O$_4$ and (**b**) without Fe$_3$O$_4$

Obviously, these are limited studies and have been specifically tailored to the application of the material of interest. Overall, it is these types of specific studies that will need to be completed to ensure that hydrogel nanocomposites can be safely used in biomedical applications.

Preliminary studies in our lab have been performed to determine the biocompatibility of a magnetic hydrogel nanocomposite composed of NIPAAm and iron oxide nanoparticles. This nanocomposite has the potential to be used in applications such as remote-controlled drug delivery and hyperthermia applications owing to its heating ability upon exposure to an alternating magnetic field. The studies performed have examined the cytotoxicity effect of the magnetic nanocomposite in direct contact with NIH 3T3 murine fibroblasts and the platelet activation on the material. As shown in Fig. 7.7 , initial results indicate that the nanocomposites do not adversely affect the fibroblasts as seen by the live cells available on the materials after 7 days.

7.5 Concluding Remarks

Hydrogel nanocomposites are novel biomaterials that can be used in a wide variety of applications including tissue engineering, drug delivery, and hyperthermia treatment. They are often advantageous to conventional hydrogels in that they provide improved properties such as increased mechanical strength and unique properties such as remote control actuation. These improvements can allow for hydrogel systems to be used in areas (e.g. bone tissue engineering) where they would not have been used before.

The safety of hydrogel nanocomposites can be evaluated through their biocompatibility responses. So far, a limited amount of research is available showing the biocompatibility of these systems, although there is a large amount of information available on the hydrogel and nanoparticle components that make up the composites. Although it would be ideal to assume that since each of the constituents of a hydrogel nanocomposite is biocompatible that the entire system would also be, however, this is not necessarily a wise assumption. It is necessary to evaluate each composite based on the specific application, it will be used in whether that is tissue engineering, biosensors, or drug delivery. Only at this point can a hydrogel nanocomposite be deemed safe for biomedical use.

References

Amarnath LP, Srinivas A, Ramamurthi A (2006) In vitro hemocompatibility testing of UV-modified hyaluonan hydrogels. Biomaterials 27:1416–1424

Bajpai AK (2007) Blood protein adsorption onto macroporous semi-interpenetrating polymer networks (IPNs) of poly(ethylene glycol) (PEG) and poly(2-hydroxyethyl methacrylate) (PHEMA) and assessment of in vitro blood compatibility. Polym Inter 56:231–244

Bajpai AK, Shrivastava M (2002) Water sorption dynamics of a binary copolymeric hydrogel of 2-hydroyethyl methacrylate (HEMA). J Biomater Sci Polymer Edn 13:237–256

Bandi S, Bell M, Schiraldi DA (2005) Temperature-responsive clay aerogel-polymer nanocomposites. Macromolecules 38:9216–9220

Barbucci R, Leione G, Magnani A, Montanaro L, Arciola CR, Peluso G, Petillo O (2002) Cu2+- and Ag+- complexes with a hyaluronane-based hydrogel. J Mater Chem 12:3084–3092

Bekiari V, Pagonis K, Bokias G, Panagiotis L (2004) Study of poly(N,N-demethacrylamide)/CdS nanocomposite organic/inorganic gels. Langmuir 20:7972–7975

Brunner TJ, Wick P, Manser P, Spohn P, Grass RN, Limbach LK, Bruinink A, Stark WJ (2006) In vitro cytotoxicity of oxide nanoparticles: comparison to asbestos, silica, and the effect of particle solubility. Environ Sci Technol 40:4374–4381

Cheng FU, Si CH, Tand YU, Yeh CS, Tsai CY, Wu CL, Wu MT, Shieh DB (2005) Characterization of aqueous dispersions of Fe_3O_4 nanoparticles and their biomedical applications. Biomater 26:729–738

Costatini A, Luciani G, Annunziata G, Silvestri B, Branda F (2006) Swelling properties and bioactivity of silica gel/pHEMA nanocomposites. J Mater Sci 17:319–325

Degirmenbasi N, Kalyon DM, Birinci E (2006) Biocomposite of nanohydroxyapatite with collagen and poly(vinyl alcohol). Colloid Surf B: Biointerfaces 48:42–49

De Groot CJ, Van Luyn MJA, Van Dijk-Wolthuis WNE, Cadee JA, Plantinga JA, Otter WD, Hennink WE (2001) In vitro biocompatibility of biodegradable dextran-based hydrogels tested with human fibroblasts. Biomater 22:1197–1203

De Queiroz AAA, Ferraz HG, Abraham GA, Fernandez MM, Bravo AL, San Roman J (2002) Development of new hydroactive dressing based on chitosan membranes: characterization and in vivo behavior. J Biomed Mat Res A 64A(1):147–154

Fenglan X, Yubao L, Xuejiang W, Jie W, Aiping Y (2004) Preparation and characterization of nano-hydroxyapatite/poly(vinyl alcohol) hydrogel biocomposite. J Mater Sci 39:5669–5672

Frimpong RA, Hilt JZ (2007) Hydrogel nanocomposites for intelligent therapeutics. In: Peppas NA, Hilt JZ, Thomas JB (eds) Nanotechnology in therapeutics: current technology and applications. Horizon Press, Norwich, pp 241–256

Frimpong RA, Fraser S, Hilt JZ (2006) Synthesis and temperature response analysis of magnetic-hydrogel nanocomposites. J Biomed Mat Res A 80A(1):1–6

Gattas-Asfura KM, Zheng Y, Micic M, Snedaker M, Ji X, Sui G, Orbulescu J, Andreoloulos FM, Pham SM, Wang C, Leblac RM (2003) Immobilization of quantum dots in the photo-cross-linked poly(ethylene glycol)-based hydrogel. J Phys Chem B 107:10464–10469

Giordano C, Sanginario V, Ambrosio L, Di Silvio L, Santin M (2006) Chemical–physical characterization and in vitro preliminary biological assessment of hyaluronic acid benzyl ester-hydroyapatite composite. J Biomater App 20:237–252

Grassmann O, Lobmann P (2004) Biomimetic nucleation and growth of $CaCO_3$ in hydrogels incorporating carboxylate groups. Biomaterials 25:277–282

Gupta AK, Curtis ASG (2004) Surface modified superparamagnetic nanoparticles for drug delivery: interaction studies with human fibroblasts in culture. J Mater Sci: Mater Med 15:493–496

Gupta AK, Gupta M (2005) Cytotoxicity suppression and cellular uptake enhancements of surface modified magnetic nanoparticles. Biomaterials 26:1565–1573

Hanks CT, Wataha JC, Sun Z (1996) In vitro models of biocompatibility: a review. Dental Mater 12:186–193

Haraguchi K, Li HJ (2005) Control of the coil-to-globule transition and ultrahigh mechanical properties of PNIPA in nanocomposite hydrogels. Angew Chem Int Ed 44:6500–6504

Haraguchi K, Takehisa T, Fan S (2002) Effects of clay content on the properties of nanocomposite hydrogels composite of poly(N-isopropylacrylamide) and clay. Macromolecules 35: 10162–10171

Haraguchi K, Farnworth R, Ohbayashi A, Takehisa T (2003) Compositional effects on mechanical properties of nanocomposite hydrogels composed of poly(N,N-dimethylacrylamide) and clay. Macromol 36:5732–5741

Haraguchi K, Takehisa T, Ebato M (2006) Control of cell cultivation and cell sheet detachment on the surface of polymer/clay nanocomposite hydrogels. Biomacromolecules 7:3267–3275

Hoffman AS (2002) Hydrogels for biomedical applications. Adv Drug Del Rev 43:3–12

Hutchens SA, Benson RS, Evans BR, O-Neill HM, Rawn CJ (2006) Biomimetic synthesis of calcium-deficient hydroxyapatite in a natural hydrogel. Biomater 27:4661–4670

Kim BC, Spinks G, Too CO, Wallace GG, Bae YH (2000) Preparation and characterization of processable conducting polymer-hydrogel composites. React Funct Polym 44:31–40

Kim DK, Zhang Y, Voit W, Rao KV, Kehr J, Bjelke B, Muhummed M (2001) Superparamagnetic iron oxide nanoparticles for bio-medical applications. Scripta Mater 44:1713–1717

Kimura M, Takai M, Ishihara K (2006) Biocompatibility and drug release behavior of spontaneously formed phospholipid polymer hydrogels. J Biomed Mater Res A 80A:45–54

Kirchner C, Liedl T, Kudera S, Pellegrino T (2005) Cytotoxicity of colloidal CdSe and CdSe/ZnS nanoparticles. Nanolett 5:331–338

Le WF, Chen YC (2006) Effects of intercalated hydrotalcite on drug release behavior for poly(acylic acid-co-N-isopropyl acrylamide)/intercalated hydrotalcite hydrogels. Europ Polym J 42:1634–1642

Lee WF, Fu YT (2002) Effect of montmorillonite on the swelling behavior and drug-release behavior of nanocomposite hydrogels. J Appl Polym Sci 89:3652–3660

Lee WF, Ju LL (2004) Effect of the intercalation agent content of montmorillonite on the swelling behavior and drug release behavior of nanocomposite hydrogels. J Appl Polym Sci 94: 74–82

Liang L, Liu J, Gong X (2000) Thermosensitive poly(N-isopropylacrylamide)-clay nanocomposite with enhanced temperature response. Langmuir 16:9895–9899

Liu TY, Hu SH, Liu KH, Liu DM, Chen SY (2006) Preparation and characterization of smart magnetic hydrogels and its use for drug release. J Magn Magn Mater 304:e397–e399

Ma J, Xu Y, Zhang Q, Zha L, Liang B (2007) Preparation and characterization of pH- and temperature-responsive semi-IPN hydrogels of carboxymethyl chitosan with poly(N-isopropyl acrylamide) crosslinked by clay. Colloi Polym Sci 285:479–484

Millon LE, Wan WK (2005) The polyvinyl alcohol-bacterial cellulose system as a new nanocomposite for biomedical applications. J Biomed Mater Res B: Appl Biomater 79:245–253

Muller K, Skepper JN, Pasfai M, Trivedi R, Howarth S, Corot C, Lancelot E, Thompson PW, Brown AP, Gillard JH (2007) Effect of ultrasmall superparamagnetic iron oxide nanoparticles (Ferumoxtran-10) on human monocyte-macrophages in vitro. Biomaterials 28:1629–1642

Nakabayashi N, Williams DF (2003) Preparation of non-thrombogenic materials using 2-methacryloyloxyethyl phosphorylcholine. Biomater 24:2431–2435

Pardo-Yissar V, Bourenko T, Wasserman J, Willner I (2002) Solvent-switchable photoelectrochemistry in the presence of CdS-nanoparticle/acrylamide hydrogels. Adv Mater 14:670–673

Peppas NA, Bures P, Leobandung W, Ichikawa H (2000) Hydrogels in pharmaceutical formulations. Eur J Pharm Biopharm 50:27–46

Peppas NA, Hilt JZ, Khademhosseini A, Langer R (2006) Hydrogels in biology and medicine: from molecular principles to bionanotechnology. Adv Mater 18:1345–1360

Pizzoferrato A, Ciapetti G, Stea S, Cenni E, Arciola CR, Granchi D, Savarino L (1994) Cell culture methods for testing biocompatibility. Clin Mater 15:173–190

Prashantha K, Rashmi BJ, Venkatesha TV, Lee JH (2006) Spectral characterization of apatite formation on poly(2-hydroxyethylmethacrylate)-TiO2 nanocomposite film prepared by sol-gel process. Spectrochimica Acta Part A 65:340–344

Prokop A, Kozlov E, Carlesso G, Davidson JM (2002) Hydrogel nanocomposite as a synthetic intro-occular lens capable of accommodation. Adv in Polym Sci 160:119–173

Ratner B, Hoffman AS, Schoen FJ, Lemons JE (2004) Biomaterials science: an introduction to materials in medicine, 2nd edn. Elsevier, San Diego, CA

Sanginario V, Ginebra MP, Tanner KE, Plannel JA, Ambrosio L (2006) Biodegradable and semi-biodegradable composite hydrogels as bone substitutes: morphology and mechanical characterization. J Mater Sci: Mater Med 17:447–454

Santiago F, Mucientes AE, Osorio M, Rivera C (2007) Preparation of composites and nanocomposite based on bentonite and poly(sodium acrylate): effect of amount of bentonite on the swelling behavior. Europ Polym J 42:1–9

Saravanan P, Raju MP, Alam S (2007) A study on synthesis and properties of Ag nanoparticles immobilized polyacrylamide hydrogel composites. Mater Chem & Phys 103:278–282

Satarkar, N, Hilt, JZ (2008) Hydrogel nanocomposites as remote controlled biomaterials. Acta Biomaterialia 4:11–16

Schiraldi C, D'Agostino A, Oliva A, Flamma F, De Rosa A, Apicella A, Aversa R, De Rosa M (2004) Development of hybrid materials based on hydroxyethylmethacrylate as supports for improving cells adhesion and proliferation. Biomaterials 25:3645–3653

Sershen SR, Westcott SL, Halas NJ, West JL (2000) Temperature-sensitive polymer-nanoshell composites for photothermally modulated drug delivery. J Biomed Mater Res 51:293–298

Sershen SR, Mensin GA, Ng M, Halas NJ, Beebe DJ, West JL (2005) Independent optical control of microfluidic valves formed from optomechanically responsive nanocomposite hydrogels. Adv Mater 17:1366–1368

Shen X, Tong H, Zhu Z, Wan P, Hu J (2007) A novel approach of homogeneous inorganic/organic composites through in situ polymerization in poly-acrylic acid gel. Mater Lett 61:629–634

Shi J, Guo ZX, Luo H, Li Y, Zhu D (2005) Actuator based on MWNT/PVA hydrogels. J Phys Chem B Lett 109:14789–14791

Shin H, Temenoff JS, Mikos AG (2003) In vitro cytotoxicity of unsaturated oligo[poly(ethylene glycol) fumarate] macromers and their cross-linked hydrogels. Biomacromolecules 4:552–560

Sinha A, Das G, Sharma BK, Roy RP, Pramanick AH, Nayar S (2007) Poly(vinyl alcohol)-hydroxyapatite biomimetic scaffold for tissue regeneration. Mater Sci Eng C 27:70–74

Smart SK, Cassady AI, Lu GQ, Martin DJ (2006) The biocompatibility of carbon nanotubes. Carbon 44:1034–1047

Sugiyami K, Matsumoto T, Yamazaki Y (2000) Evaluation of biocompatibility of the surface of polyethylene film modified with various water soluble polymers using Ar plasma-post polymerization technique. Macromol Mater Eng 282:5–12

Suzuki D, Kawaguchi H (2006) Hybrid microgels with reversible changeable multiple brilliant color. Langmuir 22:3818–3822

Tang Q, Lin J, Wu Z, Wu J, Huang M, Yang Y (2007) Preparation and photocatalytic degradability of TiO2/polyacrylamide composite. Europ Polym J 43:2214–2220

Trojani C, Boukhechba F, Scimeca JC, Vandenbos F, Michiels JF, Daculsi G, Boileau P, Weiss P, Carle GF, Rochet N (2006) Ectopic bone formation using an injectable biphasic calcium phosphate/Si-HPMC hydrogel composite loaded with undifferentiated bone marrow stromal cells. Biomaterials 27:3256–3264

Tsay JM, Michalet X (2005) New light on quantum dot cytotoxicity. Chem Biol 12:1159–1161

Tshikhudo TR, Wang Z, Brust M (2004) Biocompatible gold nanoparticles. Mater Sci Tech 20:980–984

Vihola H, Laukkanen A, Valtola L, Tenhu H, Hirvonen J (2005) Cytotoxicity of thermosensitive polymers poly(N-isopropylacrylamide), poly(N-vinylcaprolactam), and amphiphilically modified poly(N-vinylcaprolactam). Biomaterials 26:3055–3064

Wang C, Flynn NT, Langer R (2004) Controlled structure and properties of thermoresponsive nanoparticle-hydrogel composites. Adv Mater 16:1074–1079

Weian Z, Wei L, Yue'e F (2005) Synthesis and properties of novel hydrogel nanocomposites. Mater Lett 59:2876–2880

Weng H, Zhou J, Tan L, Hu Z (2004) Tissue responses to thermally-responsive hydrogel nanoparticles. J Biomater Sci Polymer Edn 15(9):1167–1180

Wick P, Manser P, Limbach LK, Dettlaff-Weglikowska U, Krumeich F, Roth S, Stark WJ, Bruinink A (2007) The degree and kind of agglomeration affect carbon nanotube toxicity. Toxic Lett 168:121–131

Williams DH (1986) Definitions in biomaterials: proceedings of a consensus conference of the european society for biomaterials, Chester, England

Xiang Y, Peng Z, Chen D (2006) A new polymer/clay nano-composite hydrogel with improved response rate and tensile mechanical properties. Europ Polym Journ 42:2125–2132

Xuli PM, Filipcsei G, Zrinyi M (2000) Preparation and responsive properties of magnetically soft poly(N-isopropylacrylamide) gels. Macromol 33:1716–1719

Yu J, Sundaram S, Weng D, Courtney JM, Moran CR, Graham NB (1991) Blood interactions with novel polyurethaneurea hydrogels. Biomaterials 12:119–120

Zhao H, Argoti SD, Farrel BP, Shipp DA (2003) Polymer-silicate nanocomposites produced by in situ atom transfer radical polymerization. J Polym Sci A: Polym Chem 42:916–924

Chapter 8
Cytotoxicity and Genotoxicity of Carbon Nanomaterials

Amanda M. Schrand, Jay Johnson, Liming Dai, Saber M. Hussain, John J. Schlager, Lin Zhu, Yiling Hong, and Eiji Ōsawa

Abstract With the recent development in nanoscience and nanotechnology, there is a pressing demand for assessment of the potential hazards of carbon nanomaterials to humans and other biological systems. This chapter summarizes our recent in vitro cytotoxicity and genotoxicity studies on carbon nanomaterials with an emphasis on carbon nanotubes and nanodiamonds. The studies summarized in this chapter demonstrate that carbon nanomaterials exhibit material-specific and cell-specific cytotoxicity with the general trend for biocompatibility: nanodiamonds > carbon black powders > multiwalled carbon nanotubes > single-walled carbon nanotubes, with macrophages being much more sensitive to the cytotoxicity of these carbon nanomaterials than neuroblastoma cells. However, the cytotoxicity to carbon nanomaterials could be tuned by functionalizing the nanomaterials with different surface groups. Multiwalled carbon nanotubes and nanodiamonds, albeit to a less extend, can accumulate in mouse embryonic stem (ES) cells to cause DNA damage through reactive oxygen species (ROS) generation and to increase the mutation frequency in mouse ES cells. These results point out the great need for careful scrutiny of the toxicity of nanomaterials at the molecular level, or genotoxicity, even for those materials like multiwalled carbon nanotubes and nanodiamonds that have been demonstrated to cause limited or no toxicity at the cellular level.

Contents

A.M. Schrand (✉)
Department of Chemical and Materials Engineering, School of Engineering and UDRI, University of Dayton, 300 College Park, Dayton, OH 45469-0240, USA

T.J. Webster (ed.), *Safety of Nanoparticles*, Nanostructure Science and Technology, DOI 10.1007/978-0-387-78608-7_8, © Springer Science+Business Media, LLC 2009

8.1 Introduction

Carbon is an essential part of living organisms and is abundant in nature in pure forms and in combination with other elements. From an engineering perspective, carbon is of great importance because of its ability to form different structures with unique properties. Carbon has long been known to exist in three forms: amorphous carbon, graphite, and diamond (Marsh 1989). The Noble-Prize-winning discovery of buckminsterfullerene C_{60} created an entirely new branch of carbon chemistry (Kroto et al. 1985; Hirsch 1994; Taylor 1995). The subsequent discovery of carbon nanotubes by Iijima opened up a new era in materials science and nanotechnology (Iijima 1991; Dresselhaus et al. 1996; Meyyappan 2005; Harris 2001). Having *conjugated* all-carbon structures with unusual molecular symmetries, fullerenes and carbon nanotubes show interesting electronic, photonic, magnetic, and mechanical properties attractive for various applications (Dai 2004, 2006). Nanodiamonds (NDs) are the most recent additions to members of the carbon family. With the diamond structure at a nanometer scale, NDs also exhibit unique mechanical and optoelectronic properties for a variety of important applications, including field emission displays (McCauley et al. 1998) and nanotribology (Erdemir et al. 1996, 1997).

   With the recent development in nanoscience and nanotechnology, there is a pressing demand for large-scale production of carbon nanomaterials for various potential applications. The number of industrial-scale facilities for the relatively low-cost production of carbon nanomaterials continues to grow. For instance, the worldwide production of single- and multi-walled carbon nanotubes alone is projected to reach over 500 tons by 2008 (Borm et al. 2006). Consequently, the exposure to carbon nanomaterials via occupational, consumer, environmental, or biomedical arenas is expected to continue to increase at a rapid rate. Although most bulk carbon materials are considered biocompatible due to their chemical inertness to cells, their reactivity and surface properties may change dramatically once carbon is reduced in size to the nano-scale. At the nanometer scale, the wave like properties of electrons inside matter and atomic interactions are influenced by the size of the material (Goldstein 1997). As a consequence, changes in the size-dependent properties (e.g., melting points, magnetic, optic, and electronic properties) may be observed even without any compositional change (Goldstein 1997). Due to the high surface-to-volume ratio associated with nanometer-sized materials, a tremendous change in chemical properties is also possible through a reduction in size (Goldstein 1997). The small size and high surface-to-volume ratio of carbon nanomaterials could also affect interactions at the cellular level, leading to enhanced permeability through the cell membrane with a profound influence on cellular dynamics (Dai 2006; Hurt et al. 2006). On the other hand, recent studies have clearly indicated great promise for carbon nanomaterials to be used as advanced drug carriers, imaging probes, or implant biomedical devices in biological systems (Dai 2006) where the use of bulk carbon films or fibers was proven to be inadequate. Therefore, it is essential to ascertain the potential hazards of carbon nanomaterials to humans and other biological systems. While the scientific community has been so far primarily focused on the potential cytotoxicity of carbon nanomaterials, albeit with somewhat conflicting results, the study on genotoxicity of carbon nanomaterials is beginning to emerge as an important research area. In this chapter, we will summarize our recent studies on the cytotoxicity and genotoxicity of carbon nanomaterials with in vitro cell culture, though references are also made to other complementary work as appropriate. Since the biological toxicity of fullerene C_{60} and its derivatives has been reviewed elsewhere (Dai 2006), we will focus on carbon nanotubes and NDs only in this article. In what follows, we will first describe a variety of methods for the assessment of cytotoxicity and genotoxicity. Then, we will summarize our studies on cytotoxicity of carbon nanotubes and NDs. Finally, some of our recent results from carbon nanotube genotoxicity will be discussed.

8.2 Assays for the Assessment of Cytotoxicity

In vitro cytotoxicity can be investigated by monitoring the decreased mitochondrial function (e.g., MTT assay for reduction of tetrazolium salt), increased mitochondrial membrane permeability (MMP), breakdown of the cellular permeability barrier (e.g., lactase dehydrogenase (LDH) assay), decreased uptake of neutral red by lysosomes (NR assay), loss of glutathione (e.g., GSH), activation of pro-inflammatory cytokines (e.g., IL-6, IL-8, TNF-α), and generation of reactive oxygen species (ROS) with the subsequent observation of cell morphology and proliferation changes.

Because of their crucial role in maintaining cellular structure and function via aerobic adenosine triphosphate (ATP) production (Hussain and Frazier 2002), mitochondria are vulnerable targets for toxic injury by a variety of compounds. The MTT assay is a colorimetric assay based on the ability of mitochondrial dehydrogenases in viable cells to reduce the yellow-colored water soluble tetrazolium salt 3-(4,5-dimethylthiazol-2-yl)-2,5-diphenyltetrazolium bromide (MTT) dye into water-insoluble purple-colored formazan crystals. These crystals are then extracted, solubilized, and homogenized from the cells with acidified isopropanol and evaluated spectrophotometrically for changes in optical density (OD) at 570–630 nm on a microplate reader (Hussain and Frazier 2002). An additional centrifugation step to remove nanomaterials from the solution before microplate reading can be performed in order to avoid direct interference of the absorption values, as is the case in some previous studies (Schrand et al. 2007). Because MTT reduction occurs only in functional mitochondria, a decrease in MTT dye reduction compared to untreated cells is an indication of reduced cell viability.

Changes in MMP can indicate the initiation of apoptosis, which is defined by the collapse in electrochemical gradient across the mitochondria called the membrane potential ($\Delta\Psi$). The insertion of apoptotic proteins or possibly the translocation of nanomaterials into the organelle can create pores, which dissipate the transmembrane potential. A cell permeable, lipophilic Mît –E–Ψ^{TM} fluorescent reagent (i.e., a fluorescent cationic dye commonly known as JC-1) can be used to monitor the permeability of the mitochondrial membrane on a fluorescent microscopy (TRITC and FITC filters). Once internalized by a health cell, the delocalized positive charge on the dye allows it to enter the negatively charged mitochondria, where it aggregates and fluoresces red. When the mitochondrial $\Delta\Psi$ collapses, however, the reagent no longer accumulates inside the mitochondria. Instead, the dye is distributed in the cytoplasm to assume a monomeric form, which fluoresces green. The ratio of red and green can then be used to estimate the degree of mitochondrial damage and apoptotic cells.

Plasma membrane integrity can be evaluated by measuring LDH leakage with testing kits, such as the CytoTox 96® Non-Radioactive Cytotoxicity Assay from Promega. This colorimetric assay allows quantitative measurements of lactate dehydrogenase (LDH), a stable cytosolic enzyme that releases upon cell lysis. The released LDH in culture supernatants can be measured with a 30-min coupled enzymatic assay involving the conversion of a tetrazolium salt into a red formazan product, which is proportional to the number of lysed cells. The absorbance is read on a microplate reader at 490 nm and compared to controls for estimating the plasma membrane damage.

Reactive oxygen species (ROS) are naturally generated by-products of cellular redox/enzymatic reactions, including mitochondrial respiration, phagocytosis, and metabolism. However, ROS can also accumulate leading to a condition known as oxidative stress, which is involved in various pathological conditions (Farber et al. 1990). Increases in intracellular ROS (oxidative stress) represent a potentially toxic insult which, if not neutralized by antioxidant defenses (e.g., glutathione and antioxidant enzymes), could lead to membrane disfunction, protein degradation, and DNA damage (Preece and Timbrell 1989; Loft and Poulsen 1999; Halliwell et al. 1992; Yu 1994; Siesjo et al. 1989). In order to test the ROS level, the cell permeable probe (i.e., 2',7'-dichlorodihdrydrofluorescein diacetate, $H_2DCF-DA$) is internalized by cells within 30 min to allow intracellular esterases to cleave the diacetate group away from H_2DCF, which is further oxidized in the cytoplasm by intracellular ROS to a fluorescent form of the probe (i.e., dichlorofluorescein, DCF). Consequently, the relative fluorescent intensity at 530 nm ($\lambda_{EX} = 485$ nm) can be measured with a fluorescent microplate reader and compared to control cells for estimating the cumulative production of ROS over a period of nanoparticle exposure.

8.3 Cytotoxicity of Carbon Nanomaterials

Although carbon nanomaterials have been examined "in vivo" within living organisms, the scientific community has been so far primarily focused on the "in vitro" cytotoxicity study in artificial environments outside living organisms (Eisenbrand et al. 2002). Generally speaking, in vitro assays consist of subcellular systems (e.g., macromolecules, organelles), cellular systems (e.g., individual cells, co-culture, barrier systems), and whole tissues (e.g., organs, slices, explants). The cytotoxicity data obtained from such in vitro systems have been used to screen, rank, and predict the acute hazards and interaction mechanisms with animals or humans. The data obtained from in vitro cytotoxicity studies have been found to be in good correlation with acute toxicity in animals and humans after studying a diverse array of chemicals (Clemedson et al. 2000). However, kinetic factors and target organ specificities often weaken the correlation. Some other limitations associated with in vitro methods include the transformation or immortalization of the cell lines that may alter the intrinsic properties of the cells; selective toxicity in which some cell types are more sensitive than others; isolation of the cells from their natural environment; and the difficulty in studying integrated groups of cells or organ systems. Therefore, in vitro studies are conducted just as a rapid and inexpensive means to uncover the underlying toxicity mechanisms of the selected chemicals without the use of animals (Castell 1997). The general mechanism of cytoxicity thus revealed can form a basis for further assessing the potential risk of nanomaterial exposure to biological systems.

As schematically shown in Fig. 8.1, some of the possible interactions characteristic of nanomaterials in the cell culture environment include, but not limited by, the binding with components of the cell culture (such as proteins or other small molecules), electrostatic attraction of positively charged nanomaterials to the cell membrane, their attachment, internalization, and possibly release. On the other hand, cells respond to toxic stress by altering their metabolic rates, cell growth, or gene transcription (Eisenbrand et al. 2002).

8.3.1 Unmodified Carbon Nanomaterials

We first utilized unmodified multiwalled carbon nanotubes (MWNTs) suspended in RPMI cell culture media containing 10% serum for investigating their cytotoxicity with a neuronal cell line, PC-12. In these studies, we used the well-known water soluble neurotoxin and cadmium oxide (CdO) as the positive control. Fine carbon black nanoparticles (CB, 20 nm) were chosen as the negative control, which have historically been used in inhalation studies as a fine particle control (Shvedova et al. 2005).Assessment of MWNT toxicity with the MTT assay showed that PC-12 cells incubated with 50–100 µg/ml of MWNTs have significantly decreased viability compared to the untreated control (Fig. 8.2A). The corresponding measurement of plasma membrane leakage with the LDH assay showed no significant increase

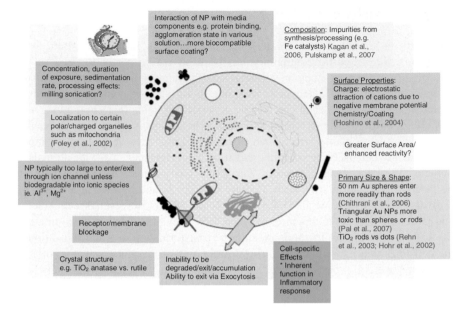

Fig. 8.1 Schematic representation of some possible interactions of nanomaterials with a cell in culture media

in leakage compared to the control at any of the nanotube concentrations tested (Fig. 8.2B).

TEM examination of cell cross sections revealed the presence of nanotube bundles both outside the cell and those beginning to penetrate the cell membrane, as marked by black arrows in Fig. 8.3. Other cell membrane features, such as finger-like projections (called filopodia as marked by white arrows), were preserved in some cells, showing the similarity in morphology to the nanotubes. It is not known whether these membrane features further prevented nanotube internalization or if

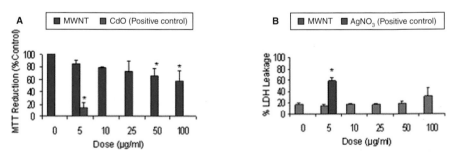

Fig. 8.2 PC-12 cell viability and membrane leakage results after incubation with 5–100 μg/ml of MWNT for 24 h. (**A**) MTT assay viability data and (**B**) LDH assay data (adapted from Schrand et al. 2007d)

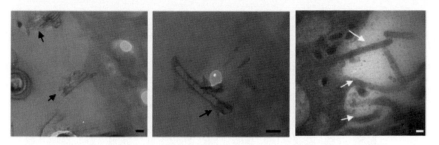

Fig. 8.3 Transmission electron microscope (TEM) images of PC-12 cells showing the interaction of MWNTs with the cell membrane (black arrows) and the similar morphology of membrane projections called filopodia (white arrows). Scale bars are 100 nm

penetration of the nanotubes into the cell membranes was merely an artifact of the processing, which depends on the aggregation behavior of MWNTs in cell culture media. The effect of nanotube dispersion on cytotoxicity will be discussed later in more details below in the section on functionalized carbon nanotubes.

8.3.2 Functionalized Carbon Nanotubes

Along with recent studies on the impact of surface functionalization on cytotoxicity (Sayes et al. 2005; Sato et al. 2005; Magrez et al. 2006), we have found that functionalization of MWNTs with sodium sulfonic acid salt ($-SO_3Na$ or -phenyl-SO_3Na) increased their biocompatibility to neuroblastoma cells with respect to unfunctionalized MWNTs or carboxylic acid ($-COOH$) functionalized MWNTs (Fig. 8.4). Screening of these different carbon nanotubes for cytotoxicity was performed with the MTT assay, which shows MWNT-SO_3Na > MWNT and SWNT-phenyl-SO_3Na > SWNT (Fig. 8.4A). The positive control CdO showed drastic decreases in viability at all concentrations in this study, indicating the validity of the assay. Another viability assay based on luminescent ATP production showed similar results, indicating that MWNT-COOH slightly reduced viability compared to CB or MWNT (Fig. 8.4B).

Other studies have shown increased biocompatibility after functionalization, but this is highly dependent on the type of functional groups introduced. For example, human dermal fibroblasts incubated with functionalized, water dispersible SWNTs (SWNT-phenyl-SO_3H, SWNT-phenyl-SO_3Na, or SWNT-phenyl-$(COOH)_2$) were found to be more biocompatible compared to unfunctionalized, surfactant stabilized SWNT (Sayes et al. 2005). However, SWNT-COOH produced a greater reduction in cell viability compared to SWNT-SO_3H. In particular, cells incubated with SWNT-COOH were significantly more toxic than the control at concentrations ranging from 10 to 200 μg/ml, whereas SWNT-phenyl-SO_3H was not toxic at concentrations up to 2 mg/ml.

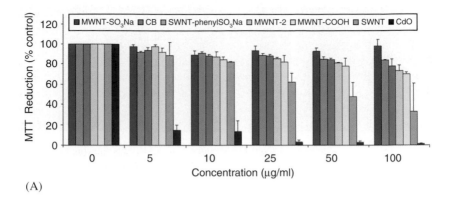

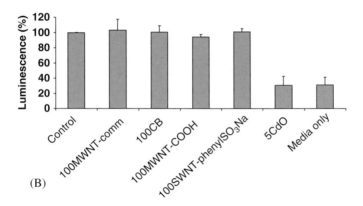

Fig. 8.4 (**A**) MTT viability assay and (ATP) assay of sodium sulfonic acid salt functionalized MWNT compared to acid functionalized MWNT-COOH or unfunctionalized MWNT in neuroblastoma cells. The first number before each of the items along the X-axis in (**B**) represents its concentration in (μg/ml)

Other notable studies have also showed that surface functionalization of carbon nanotubes with carbonyl C=O, carboxyl (COOH), and/or hydroxyl (OH) played a role in the cytotoxic response (Magrez et al. 2006; Bottini et al. 2006). Although it was shown earlier that MWNTs were the least toxic in lung tumor cells compared to CB or carbon nanofibers (CNFs), the carbon nanotubes were found to increase in cytotoxicity after acid functionalization (Magrez et al. 2006). Bottini et al. (2006) examined the toxicity of the pristine (purity > 95%) and acid-oxidized MWNTs in human T-cells at a concentration of 400 μg/ml. It was found that the pristine MWNTs reduced viability by 50% over 5 days, whereas the oxidized MWNTs reduced the viability further down to < 20%. However, the incorporation of hydrochloric acid purified MWNTs (>96 wt% purity) into polysulfone thin films was found to show good biocompatibility with both human osteoblasts and fibroblasts for up to 7 days (Chlopek et al. 2006).

The differential toxicities for functionalized and unfunctionalized carbon nanotubes may be explained by changes in surface chemistry, purity, and dispersion. For example, the dispersion of MWNT-COOH was visibly better than MWNT or CB after incubating for 24 h with neuroblastoma cells. This most probably led to a higher uptake and hence the enhanced toxicity of MWNT-COOH, as examined with the MTT and ATP assays. However, one also needs to consider possible cytotoxicity effects of the removal of amorphous carbon, residual metal catalysts, and other "contaminants" by treating the nanotube samples with certain strong acids (e.g., hydrochloric, nitric, sulfuric acid). Some chemical or physical methods used to functionalize carbon nanomaterials may also effectively prevent them from a direct contact with the cell due to changes in their surface hydrophobicity/hydrophilicity. These effects could reduce the cytotoxicity. Indeed, Wick et al. (2007) showed that their purified SWNTs dispersed well in a biocompatible surfactant and became less cytotoxic than micron-sized agglomerates of SWNTs.

8.3.3 Nanodiamonds

Since carbon nanotubes and NDs can be similarly modified for nanocomposite and biological applications (Osswald et al. 2006; Khabashesku et al. 2005), it is envisaged that ND may prove to be an excellent drug carrier, imaging probe, or implant coating in biological systems compared to currently used nanomaterials due to its optical transparency, chemical inertness, high specific area, and hardness (Huang and Chang 2004; Shenderova et al. 2002). For instance, diamond nanoparticles (NDs) have been shown to efficiently adsorb proteins due to their high surface-to-volume ratio (Huang and Chang 2004); Huang et al. 2004; Bondar et al. 2004), to readily translocate across the cell membrane due to their small size (Yu et al. 2005; Schrand et al. 2007), to alter human gene expression responsible for cancer (Bakowicz-Mitura et al. 2007), and to cause white cell destruction and erythrocyte hemolysis (Puzyr et al. 2004, Table 8.1).

Recent progresses in the dispersion of detonation NDs (2–10 nm) in aqueous media made by Ōsawa and co-workers has facilitated the use of NDs in physiological solutions (Krüger et al. 2005), whereas most previous studies have focused on polycrystalline CVD diamond films for biomedical applications (Yang et al. 2002; Poh and Loh 2004; Huang et al. 2004). Early experimental work with cells exposed to micron-sized diamond particles supports its low reactivity and high biocompatibility. For example, several studies in polymorphonuclear (PMN) leukocytes have shown that micron-sized diamond particles can be used as an inert control because they not only do not stimulate the production of ROS (Hedenborg and Klockars 1989) with no effect on degranulation or secretion of cell motility factors (Higson and Jones 1984) but also can be phagocytosed without chemotactic activity (Tse and Phelps 1970). In other cell types, such as macrophages that readily ingest large amounts of debris, micron-sized diamond dust particles were found to be non-fibrogenic (Schmidt et al. 1984). They did not affect cell viability for at least 30 h (Allison et al. 1966), and did not activate or change the cell morphology or production of interleukin 1-β (Nordsletten et al. 1996). In cells of the connective tissue and fibroblasts, micron-sized diamond particles did not induce the fibrogenic activity (Allison et al. 1966; Luhr, 1958), release of proliferation factors (Schmidt et al. 1984), or mitogenic effect (Cheung et al. 1984). In animal studies, diamond microparticles did not contribute to inflammation when introduced or injected into implant traversing canals in rabbits (Aspenberg et al. 1996), canine knee joints (Tse and Phelps 1970), or the complement system (Doherty et al. 1983). No detectable hemolysis was observed when blood was exposed to micro-sized diamond powders for 60 min (Dion et al. 1993).

More recent studies with nano-sized diamonds have also demonstrated that they are well tolerated by various cell types (Yu et al. 2005; Schrand et al. 2007). Yu et al. (2005) investigated the biocompatibility of relatively large synthetic abrasive diamond powders (100 nm) in kidney cells and found very low cytotoxicity after incubation for 3 h at concentrations up to 400 μg/ml. Since smaller detonation NDs have many advantages over other carbon nanomaterials, including its very small size, low cost synthesis, optical transparency, and ability to fluoresce after electron beam irradiation, we studied the interaction and biocompatibility of the detonation NDs (2–10 nm), with or without acid or base purification, at concentrations up to 100 μg/ml for 24 h in a variety of cell types, such as alveolar macrophages, keratinocytes, neuroblastoma cells, and PC-12 cells.

As can be seen in Fig. 8.5, neuroblastoma cells after 24 h of incubation with NDs appeared similar to controls with some elongated neurites while others remained rounded, which is typical for this heterogeneous population of cells. Agglomerates of NDs were visible in the surrounding media, at cell borders, and along neurites

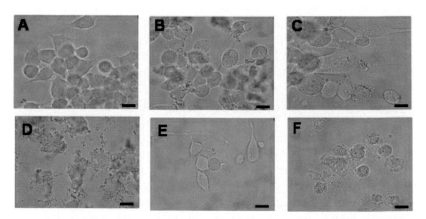

Fig. 8.5 Morphological observation of neuroblastoma cells with light microscopy after 24 h of incubation with acid or base purified nanodiamonds. (**A**) Control, (**B**) 100 μg/ml ND-raw, (**C**) 100 μg/ml ND-COOH, (**D**) 100 μg/ml ND-COONa, (**E**) 100 μg/ml ND-SO₃Na, and (**F**) 2.5 μg/ml CdO. Note that agglomerates of nanodiamonds are seen surrounding the cell borders and attached to neurite extensions whereas cell morphology is unaffected by their presence compared to the control. However, CdO induced cell shrinkage and rounding indicative of toxicity. Scale bars are 20 μm (adapted from Schrand et al. 2007b)

(Fig. 8.5B–E). In contrast, cells incubated with the positive control CdO (Fig. 8.5F) lacked cellular extensions, were reduced in size, had irregular cell borders, and formed vacuoles, which are morphological indicators of toxicity.

The internalization of NDs was further examined with transmission electron microscopy (TEM). As can be seen in Fig. 8.6, NDs were found both outside (Fig. 8.6A, white arrow) and inside the cells (Fig. 8.6A–D, black arrows) after incubation with 25 μg/ml of ND-COOH for 24 h. Higher magnification images of the selected areas in Fig. 8.6B are shown in Fig. 8.6C and D. Although these representative images do not specify the exact location of the ND nanoparticles, they appear to localize in the cytoplasm in aggregates approximately 500 nm in size and may

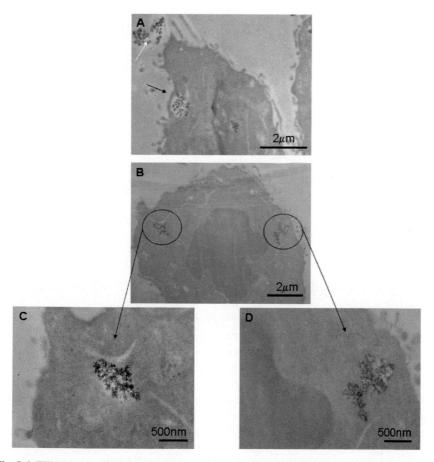

Fig. 8.6 TEM images of thin sections of neuroblastoma cells after incubation with 25 μg/ml ND-COOH for 24 h, showing ND internalization into the cytoplasm (black arrows) or NDs outside the cell (white arrow in **A**). (**C**) and (**D**) are higher magnification images of selected areas of (**B**) (adapted from Schrand et al. 2007c)

Fig. 8.7 Cytotoxicty evaluation of neuroblastoma cells after 24 h of incubation with nanodiamonds or fine carbon black nanoparticles. (**A**) MTT assay of mitochondrial function of cells indicative of cell viability and (**B**) generation of reactive oxygen species (ROS) from cells determined by the hydrolysis of DCHF-DA. Note that nanodiamonds produce no significant difference from the negative control (CB) at all concentrations tested (**A**) (adapted from Schrand et al. 2007b)

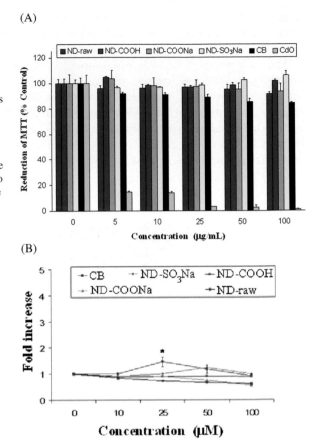

be internalized by envelopment of the plasma membrane, as indicated by the black arrow in Fig. 8.6A.

Changes in the viability of neuroblastoma cells after exposure to NDs were assessed with the MTT colorimetric assay (Fig. 8.7A). The reduction of MTT dye occurs only in functional mitochondria, therefore, a decrease in MTT dye reduction is an indication of damage to mitochondria. The positive control CdO exhibited strong toxicity with sharp decreases in cell viability compared to the negative control, CB (Fig. 8.7A). However, cells incubated with various concentrations of functionalized or unfunctionalized NDs had some slightly higher viability values, but no significant difference in viability compared to the negative control at concentrations up to 100 μg/ml (Fig. 8.7A). To further confirm the low cytotoxicity of NDs, three other cells types (macrophages, keratinocytes, and PC-12 cells) were investigated and found to display similar trends of low cytotoxicity at the same concentrations. Experiments over longer periods of time indicate that cells retain high viability for several days while incubated with NDs.

In order to investigate possible nanoparticle-induced oxidative stress, we assessed the generation of ROS (Nel et al. 2006). As mentioned earlier, the intensity of fluorescence of dichlorofluorescein (DCF), an oxidized form of 2',7'-dichlorofluorescein (DCFH), can be used as a measure of the cumulative production of ROS over a period of nanoparticle exposure. As seen in Fig. 8.7B, NDs did not initiate an oxidative stress response shown by the lack of ROS generation compared to fine carbon black nanoparticles. ROS values for cells incubated with some of the ND derivatives were even lower than the control (CB) with the exception of ND-SO$_3$Na, which was slightly greater than the control (Fig. 8.7B). The relatively low level of ROS generation produced in cells incubated with NDs is consistent with the MTT analyses described above. Therefore, these results further support the biocompatibility of NDs and suggest that ND does not induce ROS generation in this in vitro-cell model system.

Although NDs show low toxicity and no indication of oxidative stress after 24 h incubation, they may directly interact with cellular proteins of the cytoskeleton or cause changes in cell proliferation. Therefore, we examined the architecture of the actin cytoskeleton with fluorescent microscopy (Fig. 8.8). After neuroblastoma cells were incubated with 100 µg/ml of acid or base purified NDs for 24 h, distinct branching and extension of multiple neurites were found (Fig. 8.8B–E) compared to the control (Fig. 8.8A). It is currently unknown if the NDs directly bind with the actin cytoskeleton causing these alterations or if the NDs affect certain signal transduction pathways.

The above observation prompted us to further examine changes in cell growth on ND substrates. To study the behavior of neuroblastoma cells on ND substrates, cells were grown on both control (collagen or poly-L-lysine) substrates and ND-coated

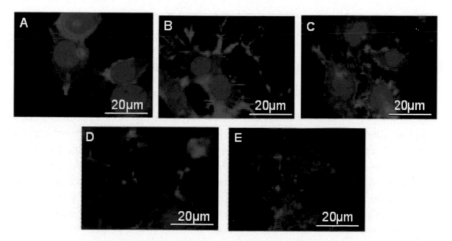

Fig. 8.8 Fluorescent microscopy of (**A–C**) mitochondrial membrane permeability assessed with Rhodamine 123 and (**D, E**) actin cytoskeleton (red) and nuclei (blue) after 24 h of incubation. (**A**) Control with no treatment, (**B**) 100 µg/ml ND-raw, (**C**) 2.5 µg/ml CdO, (**D**) Control with no treatment, (**E**) 100 µg/ml ND-raw (Schrand et al. 2007c)

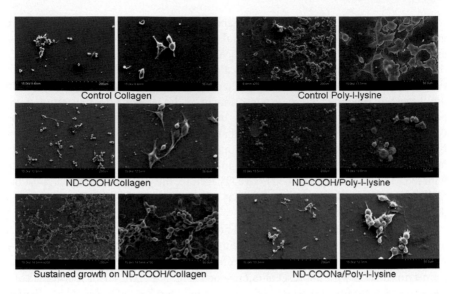

Fig. 8.9 SEM images showing the neuroblastoma cell growth on collagen or poly-L-lysine substrates coated with and without ND-COOH over 24 h (*middle row*) and 168 h (*bottom row*) (adapted from Schrand et al. 2007c)

substrates for 96 h (Fig. 8.9). Cells grown on both the control and ND-coated substrates showed attachment and neurite extension with similar morphologies visualized with scanning electron microscopy, SEM (Fig. 8.9). Cell viability and growth were examined after 168 h, showing continued trypan blue dye exclusion and visual increases in proliferation. Both suggest that these substrates would support long-term cell attachment and growth.

8.4 Differential Biocompatibility of Carbon Nanotubes and Nanodiamonds

As described above and reported elsewhere (Monteiro-Riviere and Inman 2006; Soto et al. 2005; Fiorito et al. 2006), many studies focused on the cytotoxicity and biocompatibility of carbon nanomaterials with lung or skin cells due to the risk of exposure in occupational or commercial settings. However, it is unclear whether these nanomaterials can reach the nerves associated with these organs either through internalization via the contact with skin and olfactory nerves or translocation across the blood-brain barrier. In this regard, we have also examined both neuronal (neuroblastoma) and lung (alveolar macrophage) cell lines for cytotoxicity in aqueous suspensions of carbon nanomaterials (e.g., ND, SWNT, MWNT, CB). In particular, we studied the morphological and subcellular effects of these nanomaterials on mitochondrial membrane potential and ROS generation at concentrations ranging from 25 to 100 μg/ml for 24 h incubation.

As can be seen in Fig. 8.10A, some of the neuroblastoma cells developed into their characteristic elongated extensions while others remained round upon attachment and growth of neuroblastoma in culture. After incubation with 100 μg/ml of different carbon nanomaterials (i.e. MWNTs, CBs, NDs) for 24 h, the morphologies of the cells were not noticeably different from the control cells (Fig. 8.10A). However, the nanomaterials strongly and preferentially adhere to the cell membranes in an agglomerated form even after washing the cells, as denoted by white arrows in Fig. 8.10B–D. In contrast, large agglomerates inside the cells were observed after incubating alveolar macrophages for 24 h with a lower concentration (25 μg/ml) of MWNTs or higher concentrations (100 μg/ml) of fine CBs or NDs (Fig. 8.10F–H). As denoted by black arrows, carbon nanomaterials were clearly excluded from the nucleus in macrophages while the cytoplasm was clearly filled (Fig. 8.10F–H). Because it was not obvious with light microscopy if the neuroblastoma cells had internalized the carbon nanomaterials or if they merely were attached to the cell membrane, thin sections of the cells were made and examined with TEM (Fig. 8.11), which revealed the presence of MWNTs both individually in the cytoplasm and in intracellular vacuoles.

Apart from the above morphological observation, our cytotoxicity analyses of carbon nanomaterials in both macrophages and neuroblastoma cells have revealed cell-specific and nanomaterial-specific differential toxicity. The greatest biocompatibility was found for NDs with both cell types followed the trend: ND > CB > MWNT > SWNT. For neuroblastoma cells, the interaction with SWNTs and MWNTs significantly decreased viability at concentrations from 50 to 100 μg/ml whereas CB did not decrease viability until a concentration of 100 μg/ml

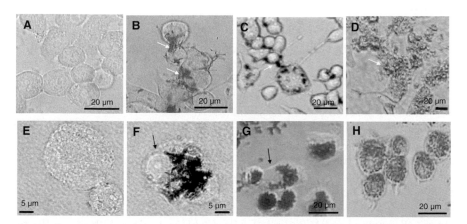

Fig. 8.10 Light microscopic examination of interactions between neuroblastoma cells or macrophages with carbon nanomaterials after 24 h incubation. (**A–D**) Neuroblastoma cells and (**E–H**) Macrophages. (**A**) Control neuroblastoma cells or cells incubated with 100 μg/ml concentrations of (**B**) MWNT, (**C**) CB, and (**D**) NDs. (**E**) Control macrophages or cells incubated with (**F**) 25 μg/ml MWNT, (**G**) 100 μg/ml CB, and (**H**) 100 μg/ml NDs. White arrows denote nanomaterials agglomerates and black arrows show the nucleus free from nanomaterials

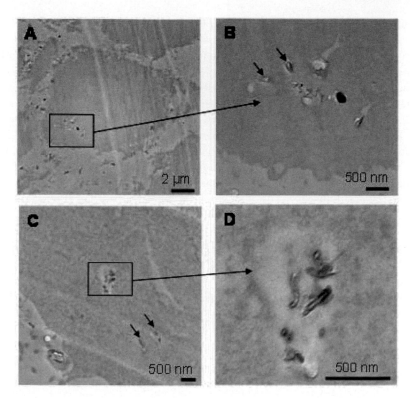

Fig. 8.11 TEM images of thin sections of neuroblastoma cells, showing MWNTs denoted by arrows within the cytoplasm and intracellular vacuoles. Boxed areas in (**A**) and (**C**) are enlarged in (**B**) and (**D**)

(Fig. 8.12A). In contrast, CdO severely reduced viability at the lowest concentration of 25 μg/ml. The viability in macrophages was significantly reduced from the control after incubation with carbon-based nanoparticles at concentrations from 25 to 100 μg/ml with the exception of ND-raw, which was not significantly different from the control at a concentration of 25 μg/ml (Fig. 8.12B).

Jia et al. (2005) compared the relative cytotoxicity of SWNTs, MWNTs, and fullerenes in macrophages. In consistent with our viability results, these authors found that SWNTs significantly impaired phagocytosis of macrophages at doses as low as 0.38 μg/ml, whereas MWNTs and C_{60} induced injury only at a high dose of 3.06 μg/ml, leading to biocompatibility in the order: $C_{60} >$ MWNTs $>$ SWNTs. In a separate, but somewhat related study, Magrez et al. (2006) found that lung tumor cells exposed to 0.02 μg/ml of CBs, CNFs, and MWNTs, respectively, for 2 days displayed increased toxicity in the order of CB $>$ CNFs $>$ MWNTs with MWNTs being the least toxic. These authors suggested that MWNTs, having the highest aspect ratio among the three carbon nanomaterials they studied, might have less dangling bonds than carbon fibers or carbon black, as the dangling bonds preferentially occur at lattice defects or endcaps.

Fig. 8.12 Various nanocarbons showing the differential toxicity due to factors such as nanomaterial functionalization, size, or shape in (**A**) neuroblastoma cells or (**B**) macrophages. Note that the similar trends for biocompatibility ND > CB > MWNT > SWNT > CdO were observed with macrophages being more sensitive to the carbon nanomaterials. Values that were significantly different from the control are denoted with asterisks (∗) (Schrand et al. 2007a)

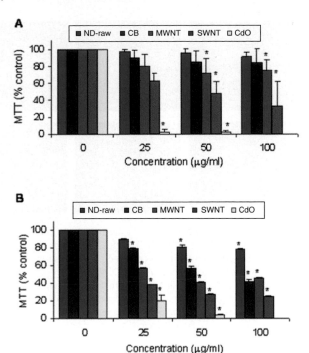

To further examine interactions between the cells and carbon nanomaterials, changes in MMP were examined with fluorescent microscopy (Fig. 8.13). The Mît –E–Ψ^{TM} fluorescent reagent when aggregated inside healthy mitochondria fluoresces red, whereas dispersion of the dye due to mitochondrial membrane leakage causes it to fluoresce green in the cytoplasm. As shown in Fig. 8.13A, B, E and F, aggregation and retention of the mitochondrial dye inside healthy cells was seen in control cells and cells incubated with 100 μg/ml ND-raw for 24 h. The dark color of the other carbon nanomaterials tended to block the fluorescent signal in certain areas compared to cells incubated with NDs, but the dispersion of the dye to the green monomeric form in the cytoplasm was apparent for CB (and MWNT, data not shown) in addition to the positive control 2.5 μg/ml CdO (Fig. 8.13C, D, G, H). This, along with the MTT viability assay, suggests that mitochondrial or apoptotic pathways may be influenced to a much greater extent by the presence of carbon nanotubes or CB compared to NDs.

Nanomaterials that generate ROS induce oxidative stress and have been linked to a general toxic response (Nel et al. 2006). We found that carbon nanotubes generated the greatest amount of ROS followed by CB, then ND in both neuroblastoma cells and macrophages (Fig. 8.14). Explanations for the decreased cell viability, increased mitochondrial membrane potential, and increased ROS after exposure to carbon nanotubes compared to CB or NDs may be attributed to the residual transition metal catalysts that were used in the synthesis of carbon nanotubes (e.g., Fe,

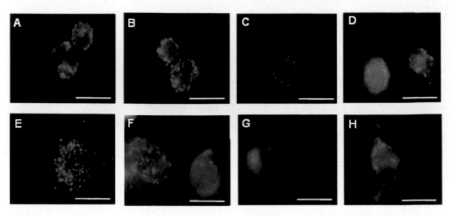

Fig. 8.13 Fluorescent microscopy of cells incubated with or without carbon nanomaterials and the positive control CdO for 24 h with the Mît-E-Ψ™ stain for mitochondrial membrane permeability detection. (**A–D**) Neuroblastoma cells and (**E–H**) Macrophages. (**A**) Control, (**B**) 100 μg/ml ND, (**C**) 100 μg/ml CB, (**D**) 2.5 μg/ml CdO, (**E**) Control, (**F**) 100 μg/ml ND, (**G**) 100 μg/ml CB, (**H**) 2.5 μg/ml CdO. Note that the control cells and cells incubated with nanodiamonds showed intact mitochondrial membranes whereas cells incubated with CB or CdO have, most likely, been damaged after the nanoparticle exposure, indicating the mitochondrial membrane leakage and the initiation of apoptosis. Scale bars are 20 μm

Ni, Co). In our studies, the estimated amount of Fe that could be extracted with nitric acid was 0.49 wt% Fe for MWNT vs. 0.26 wt% Fe for SWNT compared to levels below the detection limit for CB and NDs. The detected residual catalyst levels correspond well to the highest ROS levels in MWNTs followed by SWNTs, CB, and NDs. The main impurity in the CB was sulfur (0.43 wt%), but it is suspected that this element has a weaker link to toxicity. In support of the effect of Fe on cell viability, Garibaldi et al. (2006) found that cardiac muscle cells incubated for up to 3 days with 200 μg/ml of highly purified SWNTs were only slightly modified in shape due to SWNT binding to the cell membranes with cell viability remaining >90%. In contrast, the incubation with SWNTs retaining 26 wt% Fe (from the catalyst used for the nanotube growth) at 0.12 mg/ml SWNTs for 2 h was found to cause significant loss of glutathione (GSH) and accumulation of lipid peroxidases in macrophages, dosed with (Kagan et al. 2006), compared to iron-"free" (0.23 wt%) SWNTs under the same condition. Pulskamp et al. (2007) studied the toxic response of rat macrophages and human A549 lung cells to MWNTs, SWNTs, and acid-treated SWNTs (in addition to carbon black). In this particular case, the cells were incubated with the nanomaterials at concentrations of 5, 10, 50, and 100 μg/ml for durations of 24, 48, and 72 h. These authors found that incubation with 100 μg/ml of MWNTs for 24 h led to retention of only 20–40% viable macrophages according to the MTT assay. Additionally, all of the carbon nanotubes increased ROS production compared to the control except for the purified SWNTs, suggesting that the catalyst residues may be directly involved in the oxidative response.

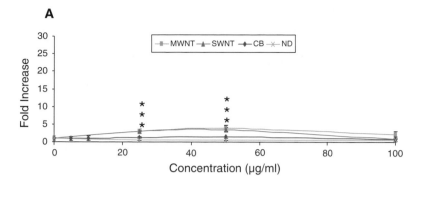

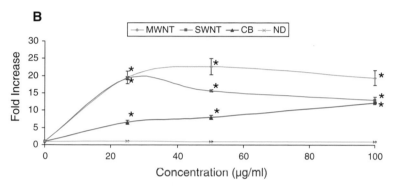

Fig. 8.14 Generation of reactive oxygen species (ROS) determined by the hydrolysis of DCFH-DA after 24 h of incubation with various carbon nanomaterials in (**A**) neuroblastoma cells and (**B**) macrophages. Note that macrophages produce approximately five times the ROS when exposed to the same nanomaterials at the same concentrations as neuroblastoma cells. All values were significantly different from the control (CB) with the exception of the NDs (**A**, **B**) and 100 μg/ml concentrations in (**A**). (Adopted from Schrand et al. 2007a)

To summarize our discussion on the cytotoxicity of carbon nanomaterials, a table with cytotoxicity data for carbon nanotubes and NDs is given in Table 8.1. This is not meant to be an exhaustive list for all of the published literature on the subject, but merely as an aid to direct the reader to the pertinent details of the articles that have been discussed.

8.5 Genotoxicity of Carbon Nanomaterials

We have thus far focused primarily on the studies of carbon nanomaterial toxicity at the cellular level. In what follows, we will discuss some of our recent results concerning possible CNT-induced DNA damage and mutagenic effects at the molecular level. This sort of information is still largely lacking in the literature, though the health effects of CNTs have attracted considerable attention. As seen above, we

Table 8.1 Selected recent studies on the cytotoxicity of carbon nanomaterial

NP type(s)/size	Properties	Cell line/animal	Concentration	Exposure time	Result (T/NT)	Reference
MWNT	Oxidized vs. Pristine (>95% purity)	T lymphocyte	400 µg/ml	5 days	Toxic	Bottini et al., 2006
MWNT	CVD synthesis, HCl purified	Human umbilical vein endothelial cells	0.5–0.9 µg/ml	120 h	Non-toxic	Flahaut et al., 2006
MWNT, SWNT	SWNT purity 90%, 95% MWNT (0.6% Ni)	Alveolar macrophage	0.38–226 µg/cm²	6 h	Toxic	Jia et al., 2005
SWNT	Iron rich 26 wt% vs. Iron-"free" 0.23 wt%	RAW 264.7 macrophages	120–500 µg/ml	up to 2.5 h	Iron Rich - Toxic	Kagan et al., 2006
MWNT	HCl for purification, H_2SO_4 for surface chemical groups C=O, COOH, OH	Lung tumor cells (H596, H446, Calu-1)	0.002–0.2 µg/ml	4 days	Toxic	Magrez et al., 2006
SWNT, MWNT-1, MWNT-2	Carbon Nanotechnologies, Inc. 5-10%Fe (SWNT)	Murine macrophage	0.005–10 µg/ml	48–54 h	Toxic	Murr et al., 2005
MWNT, SWNT, SWNT-acid treated	0.6–2.8 wt%Co (MWNT), 1.3 wt%Co/1.2 wt% Ni (SWNT), 0.6 wt%Co (purified SWNT)	Rat macrophages (NR-8383) and human lung A549 cells	5–100 µg/ml	24 h	Toxic	Pulskamp et al., 2007
SWNT	phenyl-SO_3H, phenyl-SO_3Na, phenyl-$(COOH)_2$	Human dermal fibroblasts	3 – 30,000 µg/ml	48 h	Non-toxic	Sayes et al., 2005
ND	100 nm, fluorescent	Human kidney cells (293T) Neuroblastoma, PC-12, Macrophage, Keratinocyte, PC-12	up to 400 µg/ml	3 h	Non-toxic	Yu et al., 2005
ND	2–10 nm, acid or base purified		5–100 µg/ml	24 h	Non-toxic	Schrand et al., 2007

have previously demonstrated that MWNTs generated ROS (e.g., superoxide anions, hydroxyl radicals, hydrogen peroxide) after having entered into certain mammalian cells (Schrand et al. 2007a). Free radicals can chemically alter DNA bases and cause DNA damage within the cell. We have chosen mouse embryonic stem (ES) cells as a sensitive assay to assess MWNT-mediated DNA damage, as ES cells have previously been shown to be susceptible to DNA damaging agents (Hong and Stambrook 2004; Lin et al. 2005). In response to DNA damage, eukaryotic cells, including ES cells, have developed several mechanisms to protect genomic integrity. In the presence of damaged DNA, for instance, the p53 protein is activated by protein phosphorylation as a master guardian that activates cell cycle checkpoints and triggers cell cycle arrest to provide time for the DNA damage to be repaired (Finlay et al. 1989; Bartek and Lukas 2001). Enhanced expression of p53 could also trigger cell death by apoptosis if the DNA damage is beyond repair (Sherr 2004), while under normal conditions (absence of DNA damage) p53 is expressed at low levels. The close relationship between p53 activation and DNA damage makes p53 the molecular marker of choice for assessing the genotoxicity of MWNTs to mouse ES cells. ES cells are a unique cell population with the ability to undergo both self-renewal and differentiation with the potential to give rise to all cell lineages and an entire organism (Abbondanzo et al. 1993; Odorico et al. 2001; Watt and Hogan 2000). It has been shown that ES cells are highly sensitive to DNA damaging agents (Hong and Stambrook 2004; Lin et al. 2005), and that mutant stem cells can act as the seed for cancer development. In fact, cancer is being increasingly viewed as a stem-cell disorder (Marx 2003; Reya et al. 2001). The sensitivity of ES cells to DNA damage and the importance of DNA mutations in ES cells to cancer development prompted us to study the genotoxicity of MWNTs in mouse ES cells using p53 as a molecule marker.

Figure 8.15a shows cellular uptake and response in ES cells after administering MWNTs and incubation for periods of 4 and 24 h. As can be seen, the ES cells began uptake of MWNTs after 4-h exposure and continued to accumulate throughout the time course of this study. The AP staining results showed that MWNTs reduced the red color of the AP-stained ES cells, suggesting that MWNTs reduced the stem cell marker expression. Furthermore, as shown in Fig. 8.15b, Annexin V-FITC staining result of the ES cells treated with 100 μg/ml of MWNTs for 24 h (*right image*) with respect to the untreated cells (*left image*) indicated that MWNTs induced ES cells undergo apoptosis through one of the molecular mechanisms to maintain the ES cell genomic integrity in response to DNA damage (Lin et al. 2005).

To investigate the possible molecular mechanisms causing DNA damage by MWNTs, we analyzed p53 expression levels by Western blot following the MWNT treatment. The results clearly showed that p53 protein expression level increased within 2 h after the cells were exposed to MWNTs (Fig. 8.16). Furthermore, the amount of p53 expression seemed to correlate with the amount of carbon nanotubes used in the treatment (Lanes 2 and 3). The results also indicated that p53 induction could be altered by nanoparticle surface modification. The cells that were treated with sodium hydroxide modified ND (Lane 5) had less p53 accumulation compared with the cells treated with raw NDs (Lane 4). In contrast, NDs modified with some

a.

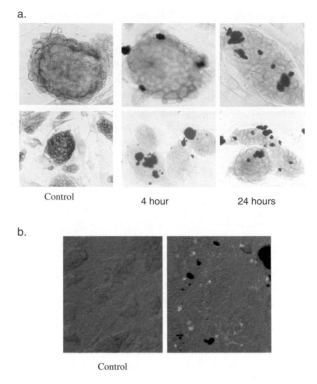

Control   4 hour   24 hours

b.

Control

Fig. 8.15 (**a**) Mouse ES cells were treated with 100 μg/mL of MWNTs for 4 and 24 h. ES cells (control and exposed) were washed with 1× PBS and fixed with 4% paraformaldehyde. AP stain results showed that some ES cells began to lose AP turning to white color compared with the untreated cells that indicated the MWNT induced the stem cell marker expression. The images were captured by inverted microscope (Olympus CK2) at 10× (*bottom row*) and 20× (*top row*) magnification via QCapture Pro Imaging Software. (**b**) Annexin V-FITC staining result of ES cells untreated and treated with 100 μg/ml of MWNTs for 24 h. ES cells were washed with binding buffer and stain with the Annexin V-FITC. Images were acquired with a Fluoview laser scanning confocal microscope mated to a Zeiss Axioplan upright microscope using 10x magnification

of the acids, containing –COOH and SO_3H_4 groups, enhanced the p53 expression (Lanes 6 and 7). As can be seen in Fig. 8.16, two bands were observed using the p53 monoclonal antibody (Chemicon). Using the phospho-specific antibody to p53-Ser-23 to examine p53 phosphorylation, we confirmed that MWNTs indeed induced p53 phosphorylation by the checkpoint protein kinase 2 (Chk2). The above observations suggest that MWNTs could cause DNA damage, as evidenced by the induction and accumulation of the p53 tumor suppressor protein. Based on these observations, we propose that p53 can be used as an early molecule marker to assess biocompatibility and toxicity of various nanomaterials (Lin et al. 2007).

To further investigate the specific DNA damage modes induced by MWNTs, we examined the expression of the key base excision repair (BER) pathway enzyme 8-oxoguanine-DNA glycosylase 1 (OGG1) by Western blot analysis following the

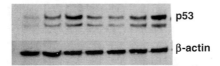

Fig. 8.16 The ES cells were lysed with RIPA buffer, and the cells lysates were analyzed by Western blot with a p53 monoclonal antibody. *Lane 1*: untreated ES cells; *Lane 2, 5* mg/ml carbon natotubes; *Lane 3*, 100 mg/ml carbon natotubes; *Lane 4*, 100 mg/ml raw nanodiamond; *Lane 5*, 100 mg/ml sodium hydroxide ONa modified nanodiamond; *Lane 6*, 100 mg/ml COOH-nanodiamonds; *Lane 7*, 100 mg/ml SO3Na-nanodiamonds. β-actin was used as a control

MWNT treatment. OGG1 is the major enzyme repairing 8-oxoguanine (8-oxoG), a mutagenic guanine base lesion produced by ROS (Hill and Evans 2006), through the removal of the oxidized bases via BER pathway (Bhakat et al. 2006). There are two isoforms of OGG1 encoded by alternatively spliced OGG1 mRNA: a 36 kDa polypeptide in the nuclear extract and a 40 kDa polypeptide in the mitochondria (Nishioka et al. 1999). As shown in Fig. 8.17, the MWNT treatment elevated the expression of both isoforms of OGG1, which suggests the occurrence of both nuclear and mitochondrial DNA damage through a mutagenic guanine base lesion.

The MWNT-induced DNA base modification described above may also cause subsequent breakdown of the DNA double strand (Ito et al. 2007). To examine this possibility, we assayed two key double strand break repair proteins: Rad51 and XRCC4 (X-ray cross-complementation group 4) involving in homologous recombination repair and non-homologous end join repair, respectively (Hoeijmakers 2001). As shown in Fig. 8.18, both proteins were up-regulated in response to the MWNT treatment.

Furthermore, Western blots revealed that the MWNT treatment increased XRCC4 expression and produced two additional higher molecular weight bands reacting with the XRCC4 antibody (Fig. 8.18b). It has been shown that in response to X-ray treatment, XRCC4 is altered by a small ubiquitin-like modifier (SUMO), which regulates its localization and function in DNA double-strand break repair (Yurchenko et al. 2006). To investigate the induction of XRCC4 Sumolation in the

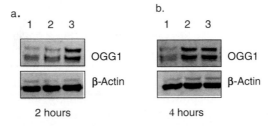

Fig. 8.17 The ES cells were lysed with RIPA buffer, and cell extracts subjected to Western blots with OGG1 polyclonal antibody. *Lane 1*, untreated; *Lanes 2 and 3*, subjected to 5 and 100 μg/mL MWNT treatment, respectively. β-actin was used as an equal loading control (adapted from Lin et al. 2007)

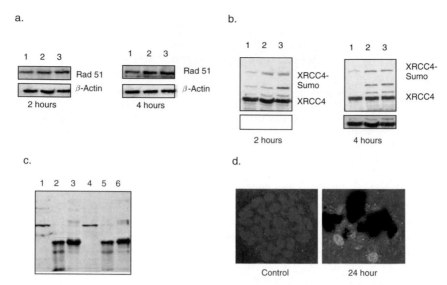

Fig. 8.18 (**a**) Cell lysates from samples treated in the same way as for Fig. 8.17 were analyzed with a Rad 51 polyclonal antibody. β-actin acted as the loading control. (**b**) Cell lysates from samples treated in the same way as for Fig. 8.17 were analyzed with a XRCC4 polyclonal antibody. β-actin was used as the loading control. (**c**) MWNT-treated cell lysates were immunoprecipitated with the XRCC4 polyclonal antibody, subjected to Western blotting, and probed with a SUMO-1 antibody. *Lanes 1–3*, untreated samples; *Lanes 4–6*, after 4 h of the MWNT treatment; *Lanes 1 and 4*, the whole cell extract; and *Lanes 2 and 5*, protein A/G PLUS-Agarose added to the cell lysate without XRCC4 antibody; and *Lanes 3 and 6*, protein A/G PLUS-Agarose added to the cell lysate with XRCC4 antibody. (**d**) Immunofluorescent staining with phosphor-specific antibody to Histone H2A-ser-139 and co-staining with Draf 5 showed that MWNTs can induce the foci formation, which indicated MWNTs caused the ES cells DNA double strand breakages. Images were acquired with a Fluoview laser scanning confocal microscope mated to a Zeiss Axioplan upright microscope using 60× oil magnification (adapted from Lin et al. 2007)

MWNT-treated mouse ES cells, we immunoprecipitated XRCC4 with the XRCC4 antibody and probed it with a SUMO-1 antibody. Figure 8.18c shows that the 95 kDa and another higher-order adduct are indeed from the SUMO-modified XRCC4, while the major band at 65 kDa corresponds to the unmodified protein as reported by Yurchenko group (Yurchenko et al. 2006). As expected, low sumoylation levels of XRCC4 were detected in the untreated cells due to the spontaneous DNA double strand breakage that occurs during the DNA replication or under tissue culture conditions. The increased level of SUMO-modified XRCC4 implicated that MWNTs indeed induced the ES cell DNA double strand breakage. This suggestion was further verified by the immunofluorescent staining with phosphor-specific antibody to Histone H2A-Ser 139 antibody (Fig. 8.18d), as Histone H2A Ser 139 phosphorylation in responses to the DNA double-strand breakage has been well characterized (Rogakou et al. 2000).

Since chemical modification of DNA bases and the breakage of DNA double strands can introduce a broad spectrum of mutations, including mitotic recombination, point mutation, and chromosome loss and translocation, we envisioned that the MWNT treatment might also increase the *mutation frequency* in ES cells. In this context, we used the endogenous molecular marker Adenine Phosphoribosyltransferase (*Aprt*) (Engle et al. 1996; Cervantes et al. 2002; Hong et al. 2006) as a reporter to examine the mutation frequency in ES cells following the MWNT treatment. The Aprt+/– heterozygous 3C4 ES cells were treated with MWNTs for 4 h. The Aprt-deficient cells can be selected in the presence of adenine analogs, such as 2-fluoroadenine (FA), whose metabolic products are cytotoxic to Aprt-proficient cells. The resistance to the adenine analog is an indication of the loss of the function of Aprt, as Aprt is known as a purine "salvage enzyme" that converts free adenine into an utilizable nucleotide (Engle et al. 1996; Cervantes et al. 2002; Hong et al. 2006). Our results indicated that the MWNT treatment increased the mutation frequency by two-fold with respect to the untreated 3C4 ES cells.

8.6 Concluding Remarks

As can be seen from above discussion, carbon nanomaterials exhibited material-specific and cell-specific cytotoxicity. Among the carbon nanomaterials and cell lines discussed in this article, a general trend for biocompatibility was observed in the order: NDs > carbon black powders > multiwalled carbon nanotubes > single-walled carbon nanotubes, with macrophages being much more sensitive to the cytotoxicity of these carbon nanomaterials than neuroblastoma cells. However, the cytotoxicity to carbon nanomaterials could be tuned by functionalizing the nano-materials with different surface groups. We have also seen that multiwalled carbon nanotubes and NDs can accumulate in mouse ES cells to cause DNA damage through ROS generation and to increase the mutation frequency in mouse ES cells. Therefore, there is a great need for careful scrutiny of the toxicity of nanomaterials at the molecular level, or genotoxicity, even for those materials like MWNTs and NDs that have been demonstrated to cause limited or no toxicity at the cellular level.

References

Abbondanzo SJ, Gadi I, Stewart CL (1993) Derivation of embryonic stem cell lines. Methods Enzymol 225:803–823

Allison AC, Harington JS, Birbeck M (1966) An examination of the cytotoxic effects of silica on macrophages. J Exp Med 124:141–154

Aspenberg P, Anttila A, Konttinen YT, Lappalainen R, Goodman SB, Nordsletten L, Santavirta S (1996) Benign response to particles of diamond and SiC: bone chamber studies of new joint replacement coating materials in rabbits. Biomaterials 17:807–812

Bakowicz-Mitura K, Bartosz G, Mitura S (2007) Influence of diamond powder particles on human gene expression. Surf Coat Technol 201:6131–6135

Bartek J, Lukas J (2001) Pathways governing G1/S transition and their response to DNA damage. FEBS Lett 490:117–122

Bhakat KK, Mokkapati SK, Boldogh I, Hazra TK, Mitra S (2006) Acetylation of human 8-oxoguanine-DNA glycosylase by p300 and its role in 8-oxoguanine repair in vivo. Mol Cell Biol 26:1654–1665

Bondar V, Pozdnyakova I, Puzyr A (2004) Applications of nanodiamonds for separation and purification of proteins. Phys Sol State 46(4):758–760

Borm PJA, Robbins D, Haubold S, Kuhlbusch T, Fissan H, Donaldson K, Schins R, Stone V, Kreyling W, Lademann J, Krutmann J, Warheit D, Oberdorster E (2006) The potential risks of nanomaterials: a review carried out for ECETOC. Particle Fibre Toxicol 3(11):1–35

Bottini M, Bruckner S, Nika K, Bottini N, Bellucci S, Magrini A, Bergamaschi A, Mustelin T (2006) Multi-walled carbon nanotubes induce T lymphocyte apoptosis. Toxicol Lett 160(2):121–126

Castell JV, Gómez-Lechón MJ (eds) (1997) In vitro methods in pharmaceutical research. Academic Press, San Diego, CA

Cervantes RB, Stringer JR, Shao C, Tischfield JA, and Stambrook PJ (2002) Embryonic stem cells and somatic cells differ in mutation frequency and type. Proc Natl Acad Sci USA 99:3586–3590

Cheung HS, Story MT, McCarty DJ (1984) Mitogenic effects of hydroxyapatite and calcium pyrophosphate dihydrate crystals on cultured mammalian cells. Arthritis Rheum 27:668–674

Chlopek J, Czajkowska B, Szaraniec B, Frackowiak E, Szostak K, Beguin F (2006) In vitro studies of carbon nanotubes biocompatibility. Carbon 44:1106–1111

Clemedson C, Barile F, Chesne C, Cottin M, Curren R, Exkwall B, Ferro M, Gomez-Lechon M, Imai K, Janus J, Kemp R, Kerszman G, Kjellstrand P, Lavrijsen K, Logemann P, McFarlane-Abdulla E, Roguet R, Segner H, Thusvander A, Walum E, Ekwall, B (2000) MEIC evaluation of acute systemic toxicity. Part VII. Prediction of human toxicity by results from testing of the first 30 reference chemicals with 27 further in vitro assays. ATLA 28:159–200

Dai L (2004) Intelligent macromolecules for smart devices: from materials synthesis to device applications. Springer-Verlag, New York

Dai L (ed) (2006) Carbon nanotechnology: recent developments in chemistry, physics, materials science and device applications. Elsevier, Amsterdam

Dion I, Roques X, Baquey C, Baudet E, Basse Cathalinat B, More N (1993) Hemocompatibility of diamond-like carbon coating. Biomed Mater Eng 3:51–55

Doherty M, Whicher JT, Dieppe PA (1983) Activation of the alternative pathway of complement by monosodium urate monohydrate crystals and other inflammatory particles. Ann Rheum Dis 42:285–291

Dresselhaus MS, Dresselhaus G, Eklund P (1996) Science of fullerenes and carbon nanotubes. Academic, San Diego, CA

Eisenbrand G, Pool-Zobel B, Baker V, Balls M, Blaauboer B, Boobis A, Carere A, Kevekordes S, Lhuguenot JC, Pieters R, Kleiner J (2002) Methods of in vitro toxicology. Food Chem Toxicol 40:193–226

Engle SJ, Stockelman MG, Chen J, Boivin G, Yum MN, Davies PM, Ying MY, Sahota A, Simmonds HA, Stambrook PJ, Tischfield JA (1996) Adenine phosphoribosyltransferase-deficient mice develop 2,8-dihydroxyadenine nephrolithiasis. Proc Natl Acad Sci USA 93:5307–5312

Erdemir A, Bindal C, Fenske GR, Zuiker C, Krauss AR, Gruen DM (1996) Friction and wear properties of smooth diamond films grown in fullerene + argon plasmas. Diamond Relat Mater 5:923–931

Erdemir A, Halter M, Fenske GR, Zuiker C, Csencsits R, Krauss AR, Gruen DM (1997) Friction and wear mechanisms of smooth diamond films during sliding in air and dry nitrogen. Tribol Trans 40:667–675

Farber JL, Kyle ME, Coleman JB (1990) Mechanisms of cell injury by activated oxygen species. J Lab Invest 62:670–679

Finlay CA, Hinds PW, Levine AJ (1989) The p53 proto-oncogene can act as a suppressor of transformation. Cell 57:1083–1093

Fiorito S, Serafino A, Andreola F, Bernier P (2006) Effects of fullerenes and single-wall carbon nanotubes on murine and human macrophages. Carbon 44:1100–1105

Garibaldi S, Brunelli C, BavastrelloV, GhigliottiG, Nicolini C (2006) Carbon nanotube biocompatibility with cardiac muscle cells. Nanotechnology 17:391–397

Goldstein AN (ed) (1997) Handbook of nanophase materials. Marcel Dekker, Inc., New York

Halliwell B, Gutteridge J, Cross CJ (1992) Free radicals, antioxidants and human disease: where are we now? J Lab Clin Med 119:598–620

Harris PJF (2001) Carbon nanotubes and related structures – new materials for the twenty-first century. Cambridge University Press, Cambridge

Hedenborg M, Klockars M (1989) Quartz dust-induced production of reactive oxygen metabolites by human granulocytes. Lung 167:23–32

Higson FK, Jones OT (1984) Oxygen radical production by horse and pig neutrophils induced by a range of crystals. J Rheumatol 11:735–740

Hill JW, Evans MK (2006) Dimerization and opposite base-dependent catalytic impairment of polymorphic S326C OGG1 glycosylase. Nucleic Acids Res. 34:1620–1632

Hirsch A (1994) The chemistry of the fullerenes. Thieme, Stuttgart

Hoeijmakers JH (2001) Genome maintenance mechanisms for preventing cancer. Nature 411: 366–374

Hong YL, Stambrook PJ (2004) Restoration of an absent G_1 arrest and protection from apoptosis in embryonic stem cells after ionizing radiation. Proc Natl Acad Sci USA 101: 14443–14448

Hong YL, Cervantes RB, Stambrook PJ (2006) DNA damage response and mutagenesis in mouse embryonic stem cells. Methods Mol Biol 329:313–326

Huang LCL, Chang H (2004) Adsorption and immobilization of cytochrome c on nanodiamonds. Langmuir 20:5879–5884

Huang TS, Tzeng Y, Liu Y, Chen Y, Walker K, Guntupalli R, Liu C (2004) Immobilization of antibodies and bacterial binding on nanodiamond and carbon nanotubes for biosensor applications. Diamond Rel Mat 13:1098–1102

Hurt RH, Monthioux M, Kane A (2006) Toxicology of carbon nanomaterials: status, trends, and perspectives on the special issue. Carbon 44:1028–1033.

Hussain S and Frazier J (2002) Cellular toxicity of hydrazine in primary hepatocytes. Tox Sci 69:424–432

Iijima S (1991) Helical microtubules of graphitic carbon. Nature 354:56–58

Ito K, Keiyo Takubo K, Arai F, Satoh H, Matsuoka S, Ohmura M, Naka K, Azuma M, Miyamoto K, Hosokawa K, Ikeda Y, Mak TW, Suda T, Hirao A (2007) Regulation of reactive oxygen species by *Atm* is essential for proper response to DNA double-strand breaks in lymphocytes. J Immunol 178:103–110

Jia G, Wang H, Yan L, Wang X, Pei R, Yan T, Zhao Y, Guo X (2005) Cytotoxicity of carbon nanomaterials: single-wall nanotube, multi-wall nanotube, and fullerene. Environ Sci Technol 39(5):1378–1383

Kagan VE, Tyurina YY, Tyurin VA, Konduru NV, Potapovich AI, Osipov AN, Kisin ER, Schwegler-Berry D, Mercer R, Castranova V, Shvedova AA (2006) Direct and indirect effects of single walled carbon nanotubes on RAW 264.7 macrophages: role of iron. Toxicol Lett 165:88–100

Khabashesku VN, Margrave JL, Barrera EV (2005) Functionalized carbon nanotubes and nanodiamonds for engineering and biomedical applications. Diamond Rel Mat 14:859–866

Kroto HW, Heath JR, O'Brien SC, Curl RF, Smalley RE (1985) C60: Buckminsterfullerene. Nature 318:162–163

Krüger A, Kataoka F, Ozawa M, Fujino T, Suzuki Y, Aleksenskii A, Vul' A, Ōsawa E (2005) Unusually tight aggregation in detonation nanodiamond: identification and disintegration. Carbon 43:1722–1730

Lin T, Chao C, Saito S, Mazur SJ, Murphy ME, Appella E, Xu Y (2005) p53 induces differentiation of mouse embryonic stem cells by suppressing Nanog expression. Nat Cell Biol 7:165–171

Lin Z, Chang DW, Dai L, Hong YL (2007) DNA damage induced by multiwalled carbon nanotubes in mouse embryonic stem cells 7(12):3592–3597

Loft S, Poulsen H (1999) Markers of oxidative damage to DNA: antioxidants and molecular damage. Methods Enzymol 300:166–184

Luhr HG (1958) Comparative studies on phagocytosis of coal powders of various carbonification grades, also of quartz and diamond powders in tissue cultures. Arch Gewerbepath 16: 355–374

Magrez A, Kasas S, Salicio V, Pasquier N, Seo J, Celio M, Catsicas S, Schwaller B, Forro L (2006) Cellular toxicity of carbon-based nanomaterials. Nano Lett 6:6:1121–1125

Marsh H (1989) Introduction to carbon science. Butterworths, London

Marx J (2003) Mutant stem cells may seed cancer. Science 301:1308–1310

McCauley TG, Corrigan TD, Krauss AR, Auciello O, Zhou D, Gruen DM, Temple D, Chang RPH, English S, Nemanich RJ (1998) Electron emission properties of Si field emitter arrays coated with nanocrystalline diamond from fullerene precursors. Mater Res Soc (MRS) Symp Proc 498:227–232

Meyyappan M (2005) Carbon nanotubes: science and applications. CRC Press, Boca Raton, FL

Monteiro-Riviere N, Inman A (2006) Challenges for assessing carbon nanomaterial toxicity to the skin.Carbon 44:1070–1078

Nel A, Xia T, Madler L, Li N (2006) Toxic potential of materials at the nanolevel. Science 311: 622–627

Nishioka K, Ohtsubo T, Oda H, Fujiwara T, Kang D, Sugimachi K, Nakabeppu Y. (1999) Expression and differential intracellular localization of two major forms of human 8-oxoguanine DNA glycosylase encoded by alternatively spliced OGG1 mRNAs. Mol Biol Cell 10:1637–1652

Nordsletten L, Hogasen AK, Konttinen YT, Santavirta S, Aspenberg P, Aasen AO (1996) Human monocytes stimulation by particles of hydroxyapatite, silicon carbide and diamond: in vitro studies of new prosthesis coatings. Biomaterials 17:1521–1527

Odorico JS, Kaufman DS, Thomson JA (2001) Multilineage differentiation from human embryonic stem cell lines. Stem Cells 19:193–204

Osswald S, Yushin G, Mochalin V, Kucheyev SO, Gogotsi Y (2006) Control of sp2/sp3 carbon ratio and surface chemistry of nanodiamond powders by selective oxidation in air. J Am Chem Soc 128:11635–11642

Poh W, Loh K (2004) Biosensing properties of diamond and carbon nanotubes. Langmuir 20: 5484–5492

Preece N, Timbrell J (1989) Investigation of lipid peroxidation induced by hydrazine compounds in vivo in the rat. Pharmacol Toxicol 64:282–285

Pulskamp K, Diabaté S, Krug HF (2007) Carbon nanotubes show no sign of acute toxicity but induce intracellular reactive oxygen species in dependence on contaminants. Tox Lett 168: 58–74

Puzyr AP, Neshumayev DA, Tarskikh SV, Makarskaya GV, Dolmatov VYu, Bondar VS (2004) Destruction of human blood cells in interaction with detonation nanodiamonds in experiments in vitro. Diamond Rel Mat 13:2020–2023

Reya T, Morrison SJ, Clarke MF, Weissman IL (2001) Stem cells, cancer, and cancer stem cells. Nature 414:105–111

Rogakou EP, Nieves-Neira W, Boon C, Pommier Y, Bonner WM (2000) Initiation of DNA fragmentation during apoptosis induces phosphorylation of H2AX histone at serine 139. J Bio Chem 275:9390–9390

Sato Y, Shibata KI, Kataoka H, Ogino SI, Bunshi F, Yokoyama A, Tamura K, Akasaka T, Uo M, Motomiya K, Jeyadevan B, Hatakeyama R, Watari F, Tohji K (2005) Strict preparation and evaulation of water-soluble hat-stacked carbon nanofibers for biomedical application and their high biocompatibility: influence of nanofiber surface functional groups on cytotoxicity. Mol BioSyst 1:142–145

Sayes C, Liang F, Hudson J, Mendez J, Guo W, Beach J, Moore V, Doyle C, West J, Billups W, Ausman K, Colvin V (2005). Functionalization density dependence of single-walled carbon nanotubes cytotoxicity in vitro. Toxicol Lett 161(2):35–142

Schmidt JA, Oliver CN, Lepe-Zuniga JL, Green I, Gery I (1984) Silica-stimulated monocytes release fibroblast proliferation factors identical to interleukin 1. A potential role for interleukin 1 in the pathogenesis of silicosis. J Clin Invest 73:1462–1472

Schrand AM, Dai L, Schlager JJ, Hussain SM, Ōsawa E (2007a) Differential biocompatibility of carbon nanotubes and nanodiamonds. Diamond Rel Mat (In Press)

Schrand AM, Huang H, Carlson C, Schlager JJ, Ōsawa E, Hussain SM, and Dai L (2007b) Are diamond nanoparticles cytotoxic? J Phys Chem B 111(1):2–7

Schrand AM, Huang H, Qu L, Schlager JJ, Ōsawa E, Hussain SM, Dai L (2007c) In vitrobiocompatibility of diamond nanoparticles with cells. SAMPE J (Submitted for publication)

Schrand AM, Kathy Szcublewski K, Schlager JJ, Dai L, Hussain SM (2007d) Interaction and biocompatibility of multi-walled carbon nanotubes in PC-12 Cells. J Neuroprotection Neuroregeneration 3(2):115–121

Shenderova O, Zhirnov V, Brenner D (2002) Carbon nanostructures. Crit Rev Solid State Mat Sci 27:227–356

Sherr CJ (2004) Principles of tumor suppression. Cell 116:235–246

Shvedova AA, Kisin ER, Mercer R, Murray AR, Johnson VJ, Potapovich AI, Tyurina YY, Gorelik O, Arepalli S, Schwegler-Berry D, Hubbs AF, Antonini J, Evans DE, Ku B, Ramsey D, Maynard A, Kagan VE, Castranova V, Baron P (2005) Unusual inflammatory and fibrogenic pulmonary responses to single-walled carbon nanotubes in mice. Am J Physiol-Lung Cell Mol Physiol 289(5):L698–L708

Siesjo B, Agardh C, Bengtsson F (1989) Free radicals and brain damage. Cerebrovasc Brain Metab Rev 1:165–211

Soto K, Carrasco A, Powell T, Garza K, Murr L (2005) Comparative in vitro cytotoxicity assessment of some manufactured nanoparticulate materials characterized by transmission electron microscopy. J Nanopart Res 7:145–169

Taylor R (ed) (1995) The chemistry of fullerenes. World Scientific, Singapore

Tse RL, Phelps P (1970) Polymorphonuclear leukocyte motility in vitro. V. Release of chemotactic activity following phagocytosis of calcium pyrophosphate crystals, diamond dust, and urate crystals. J Lab Clin Med 76:403–415

Watt FM, Hogan BLM (2000) Out of Eden: stem cells and their niches. Science 287:1427–1430

Wick P, Manser P, Limbach LK, Dettlaff-Weglikowska U, Krumeidch F, Roth S, Stark WJ, Bruinink A (2007) The degree and kind of agglomeration affect carbon nanotube cytotoxicity. Toxicol Lett 168:121–131

Yang W, Auciello O, Butler J, Cai W, Carlisle J, Gerbi J, Gruen D, Knickerbocker T, Lasseter T, Russell J, Smith L, Hamers R (2002) DNA-modified nanocrystalline diamond thin-films as stable, biologically active substrates. Nat Mat 1:253–257

Yu B (1994) Cellular defenses against damage from reactive oxygen species. Physiol Rev 74: 139–162

Yu S, Kang M, Chang H, Chen K, Yu Y (2005). Bright fluorescent nanodiamonds: no photobleaching and low cytotoxicity. J Am Chem Soc 127: 17604–17605

Yurchenko V, Xue Z, Sadofsky MJ (2006) SUMO modification of human XRCC4 regulates its localization and function in DNA double-strand break repair. Mol Cell Biol 26:1786–1794

Chapter 9
Calcium Phosphate Nanoparticles: Toxicology and Lymph Node Targeting for Cancer Metastasis Prevention

Rajesh A. Pareta

Abstract Applications of nanoparticles in biology are rapidly developing areas in nanomedicine. In cancer therapy, nanoparticles are being used for the detection, diagnosis, and imaging of tumors. Calcium phosphate has long been used as a bone substitute biomaterial and is FDA approved. It is biocompatible, easy to synthesize and relatively cheap. Due to these favorable conditions, it has been investigated for numerous drug delivery of anti-tumor drug applications. Tumor metastases are a major health issue before and after surgical removal of a tumor. Presently, chemotherapy is being used to tackle metastases, but an anti-tumor drug does not differentiate between healthy tissue and tumors. It simply kills all cells it interacts with, and hence there is a need to develop an effective targeted drug delivery system which only releases the drug in tumor cell conditions. Anti-tumor drugs can be easily adsorbed onto the surface of calcium phosphates through electrostatic interactions. Calcium phosphate nanoparticles dissolve in the acidic micro-environment of the tumor, thus, releasing the drug which eventually kills the tumor cells. Most importantly, the calcium phosphate nanoparticles can be targeted to the sentinel lymph node where they can probe tumor metastases. This chapter will cover the history and future promises of using calcium phosphate nanoparticles to treat cancer, including its safety and toxicity in such applications.

Contents

R.A. Pareta (✉)
Division of Engineering, Brown University, 182 Hope Street, Providence, RI 02912
e-mail: Rajesh_Pareta@brown.edu

T.J. Webster (ed.), *Safety of Nanoparticles*, Nanostructure Science and Technology,
DOI 10.1007/978-0-387-78608-7_9, © Springer Science+Business Media, LLC 2009

9.1 Introduction

Within medical practice, there have been efforts to achieve selective delivery of drugs specific to areas in the body in order to maximize drug action and minimize side effects. Drugs used in cancer chemotherapy represent a clear example for such cases; cytotoxic compounds not only kill target tumor cells but also normal cells in the body.

The lymphatics serve as a primary route for dissemination of many solid tumors, particularly those of epithelial origin including breast, colon, lung, and prostate. The lymphatics show many advantages over the blood circulation as a transport route for a metastasizing tumor cell or embolism, just as it does for white blood cells. The smallest lymphatic vessels are still much larger than blood capillaries, and flow velocities are orders of magnitude slower. Lymph fluid is nearly identical to interstitial fluid and promotes cell viability. Furthermore, lymph nodes provide ideal cell incubators with long residence time, areas of flow stagnation, and access to the blood stream. Because of its important role in the dissemination of some of the deadliest cancers, the lymphatic route shows great potential for targeted drug delivery to lymph node metastases (Swartz 2001).

9.2 Lymphatic System

The lymphatic system is composed of lymph, lymphatic vessels, and lymph nodes. Its essential functions are to ensure homeostasis by draining interstitial tissue to provide maintenance of interstitial pressure and plasma volume, and to reabsorb macromolecules into the circulation (Swartz and Skobe 2001). The lymphatic system also plays an essential role in immune surveillance, as lymph nodes act as filters for pathogens, and lymphatic vessels ensure the transport of immune cells to their site of action. Apart from its role in the elimination of certain antigens, the lymphatic system is also involved in the elimination of cellular debris and metabolic waste. Finally, the lymphatic system ensures transport of lipids to the liver in the form of chylomicrons following their absorption in the small intestine.

Lymph circulation inside lymphatic vessels is unidirectional from the tissues towards the lymph nodes, usually in parallel with blood vessels. Although lymphatic vessels generally have a larger diameter than contiguous blood vessels, the pressure inside lymphatic vessels is lower, and lymph circulation is slower. Truncated conical valves constitute an anti-reflux system, and the circulation of lymph in small caliber vessels is mainly ensured by peripheral muscle contractions. The lymphatic drainage of the body terminates in two collecting ducts, the left thoracic duct on the left and the right lymphatic duct on the right, which then enter the venous system at the jugulo-subclavian junction (Jussila and Alitalo 2002). Lymphatic capillaries are vessels with a diameter between 20 and 30 μm – i.e., larger than blood capillaries. Using an electron microscope, lymphatic capillaries can be distinguished from blood capillaries by their large, irregular lumen and the overlapping arrangement of lymphatic endothelial cells with a scant, vesicle-rich cytoplasm and a plasma membrane presenting numerous invaginations. Finally, lymphatic capillaries are characterized by the almost complete absence of basement membrane and pericytes, rare tight junctions, the absence of intraluminal red blood cells, but the presence of anchoring filaments (Alitalo et al. 2005).

The lymph node is an organ vascularized by an afferent artery and an efferent vein, and a large number of venules called high-endothelial venules. The cortex contains follicles and medullary cords. A capsule that gives rise to connective tissue trabecula encloses the lymph node. Afferent lymphatic vessels enter the capsule to drain the subcapsular sinus. The subcapsular sinus lymph is distributed by a series of anastomotic ducts, the medullary sinuses, in the hilum of the lymph node, which give rise to one or several efferent lymphatic vessels (Jussila and Alitalo 2002).

9.2.1 Targeting to Lymph Nodes

This system plays an important role in helping to defend the tissues against infection by filtering particles from the lymph and by supporting the activities of the lymphocytes, which furnish immunity, or resistance, to the specific disease causing agents. Also, it is well known that the lymphatic absorption of a drug after intestinal administration provides an advantage over the portal blood route for the possible avoidance of liver pre-systemic metabolism (hepatic first-pass effect). Due to such fundamental functions or characteristics, many attempts have been made to utilize the lymphatic system for the route of drug delivery, which have been reviewed by Muranishi (1991). Research into lymphatic targeting has recently attracted increasing interest not only for providing a preferential anticancer chemotherapy, but also for improving oral absorption of macromolecule drugs, or achieving mucosal immunity.

The lymphatic system is the site of many diseases such as metastitial tuberculosis, cancer, and filariasis (Weinstein 1983). Due to the peculiar nature and anatomy of the lymphatic system, localization of drugs in the lymphatics has been particularly difficult to achieve. Much effort for the lymphatic targeting of drugs has been directed towards the use of anticancer agents. Using various routes of administration such as intramuscular (Hashida et al. 1997), subcutaneous and intraperitoneal (Nakamoto et al. 1975; Parker et al. 1982), significant enhancement of drugs delivery has been reported.

9.2.2 Detecting the Lymph Nodes

Besides chemotherapeutic purposes, there has been much research in the past 30 years on the staining of lymph nodes before surgery. The retroperitoneal lymph nodes, for example, are the primary filters of metastases spreading from malignant tumors of pelvic organs. The inaccessibility of these retroperitoneal lymph nodes for therapy is still one of the many unsolved clinical problems. The staining of lymph nodes prior to surgery would improve the radicality and selectivity of lymphonodectomy because lymph nodes are very often difficult to locate since they are covered by fat tissue of similar color and consistency. Here we will discuss two methods which are most used at the moment to visualize lymph nodes.

9.2.2.1 Tracer

During this procedure, a radioactive tracer is injected around the tumor, just under the skin, and gradually migrates to the sentinel nodes. Using a gamma camera, images are taken of the area to show the location of the nodes. A blue dye is then injected near the tumor. The dye colors the lymph vessels and sentinel nodes a bright blue color that is easily seen after the skin incision is made.

9.2.2.2 Magnetic Nanoparticles

These nanoparticles target normal tissue (in particular, macrophages, a class of immune cells that swarm through healthy lymph nodes). Once taken up by macrophages, the particles interfere with the cells' magnetic properties, causing a reduction in signal, and hence, a dimming of the lymph node when viewed by magnetic resonance imaging techniques. During metastasis, tumor cells fill the lymphatic ducts, essentially preventing the flow of macrophages (and of image-dimming particles) into and out of the lymph nodes. When the particles do not get to the lymph nodes, or the lymph node is replaced by tumors, there is no decrease in signal intensity and the lymph nodes stand out as bright spots.

Lastly, Bulte (2005) injected superparamagnetic iron oxide nanoparticles intravenously. These particles were rapidly taken up by phagocytic cells and were visualized through magneto resonance to locate the lymph nodes.

9.2.3 Size Effect

The decisive factor which influences lymphatic nanoparticle disposition subcutaneously when administered. Small particles are taken up from the site of injection to a great extent, while larger particles predominately remain at the injection site. Other factors such as composition, surface charge, and surface coatings have been reported to not substantially affect the uptake (Oussoren and Storm 1998).

Reddy et al. (2006) prepared nanoparticles of various sizes (20, 45, and 100 nm) and studied the lymphatic uptake and retention. They found that 20 nm particles were most readily taken up into lymphatics following an interstitial injection, while both 45 and 100 nm showed significant retention in lymph nodes (not for 20 nm particles).

9.2.4 Lymph Node Metastases

In the natural progression of solid tumors, tumor dissemination occurs via blood vessels, lymphatic vessels, and sometimes via a cavity or along a surface (e.g., the pleural or peritoneal cavity) (Fisher and Fisher 1966).

Lymphogenous metastasis consists morphologically of intratumoral or perimoral lymphatic emboli and/or lymph node invasion, and clinically of the presence of one or several enlarged lymph nodes. The histological features of lymph node invasion are varied: tumor emboli in the subcapsular sinus, isolated tumor cell or micrometastasis within the lymph node parenchyma, or even sometimes clinically obvious lymph node metastasis, that may mimic the appearance of the primary tumor with its contingent of stromal cells and/or tumor necrosis. At an even more advanced stage, capsular effraction with invasion of the adjacent connective tissue may be observed.

It is usually considered that lymph node metastasis can only occur via peritumoral lymphatics, as intratumoral lymphatics may not be functional due to the high interstitial pressure within the tumor (Pepper 2001). However, some studies have demonstrated a relationship between the presence of intratumoral lymphatics, lymph node invasion, and poor prognosis in patients operated for squamous cell carcinoma of the head and neck (Maula et al. 2003), melanoma (Dadras et al. 2003), and papillary thyroid carcinoma (Hall et al. 2003). Similarly, although the majority of studies report the non-functional nature of intratumoral lymphatics, other studies have demonstrated the presence of cycling lymphatic endothelial cells and tumor emboli in these vessels (Kyzas et al. 2005). These discordant appearances probably depend on the tumor model studied.

Although lymph node invasion is a major factor of poor prognosis in many solid tumors (breast cancer, squamous cell carcinoma of the head and neck, melanoma, NSCLC, etc.), lymph node metastases per se are almost never life-threatening, in contrast with visceral metastases. It is therefore important to determine whether the presence of lymph node metastasis simply reflects the capacities already acquired by tumor cells to disseminate via other pathways such as hematogenous dissemination, or whether a large proportion of visceral metastases are derived from tumor cells filtered by local lymph nodes (Thiele and Sleeman 2006). The invasive capacities acquired by tumor cells facilitate their penetration into both blood vessels and lymphatics. Blood vessels constitute a high pressure system, and it has been demonstrated that only a very small proportion of tumor cells disseminated by the hematogenous route are viable and able to form a metastasis at a distant site away from the primary tumor. Inversely, the lymphatic vascular system is a low pressure system and the diameter of lymphatic vessels is greater than that of contiguous blood vessels, which could facilitate the formation of metastases in lymph nodes, which constitute a real filter allowing the selection and expansion of tumor clones with a high metastatic potential.

9.2.5 Strategies to Block Lymph Node Metastases

Due to their high rate of proliferation, tumor endothelial cells reveal susceptibility against cytotoxic drugs (Denekamp 1999). Furthermore, tumor endothelial cells are considered to be genetically stable and, thus, less prone to drug resistance. Selective delivery of cytotoxic drugs to tumor endothelium may be an approach to establish vascular targeting for tumor therapy.

In cancer chemotherapy, increased attention is currently focused on the development of drug delivery systems for systemic application with the aim of enhancing the selectivity of agents by promoting their accumulation at the site of disease. In vivo, the endothelial wall prevents many systemically circulating agents with in vitro activity from reaching the tumor cell at adequate concentrations.

The major purpose of lymphatic targeting is to provide an effective anticancer chemotherapy to prevent the metastasis of tumor cells by accumulating the drug in the regional lymph node via subcutaneous administration. The objectives of lymph targeting also involve the localization of diagnostic agents to the regional lymph node to visualize the lymphatic vessels before surgery, and the improvement of peroral bioavailability of macromolecular drugs, like polypeptides or proteins, which are known to be selectively taken up from the Peyer's patch in the intestine.

Targeted delivery of drugs can be achieved utilizing carriers with a specified affinity to the target tissue. There are two approaches for the targeting, i.e., chemical modification of drugs and pharmaceutical modification. In the case of the chemical approaches representing so-called prodrugs, the drug has to possess a suitable functional group in its molecular structure, and the method for synthesizing has to be individually developed for each drug substance. On the contrary, the pharmaceutical approach utilizing particulate carriers has such advantages that the technology, once achieved, is principally applicable to any drug, and the process is comparatively easy, which forms the basis of this chapter.

9.2.5.1 Calcium Phosphates (CaP)

Drug release from such CaP solid formulations depends on the solubilization of the drug in body fluids, adsorption to CaP, and diffusional gradients. Particulate systems can simply release drugs in the presence of chloride ions present physiologically. Barroug et al. prepared CaP nanoparticles adsorbed with cis-diamminedichloropatinum or cisplatin (CDDP) and conducted in vitro studies. The CDDP was adsorbed through electrostatic attraction on CaP nanoparticles. Less crystalline CaP released CDDP more slowly than the amorphous CaP. CDDP drug activity was retained (Barroug et al. 2004).

Tahara and Ishii (2002) concluded that HAP containing 10% CDDP (weight %) was the ideal implant for rabbits, because it achieved sustained drug release and new bone formation. The 5% and 10% implants had lower sustained release rates than the 20% implant, and the effect of the 5% and 10% implants on the kidney and liver was less significant than that of the systemic administration of the 20% implant. Bone was not formed in rabbits with the 20% implant, even by week 12. New bone was formed by weeks 12 and 6 with the 10% and 5% implants, respectively. This suggests that high concentrations of CDDP inhibited bone formation. However, the effects of different CDDP concentrations have not yet been elucidated. The present study demonstrated that, in order to achieve early bone formation, the CDDP content in an implant should not exceed 10% (w/w).

Similarly, Lebugle et al. showed that it is possible to obtain a slow release of MTX by its incorporation into a biodegradable calcium phosphate matrix, which is well known for its biocompatibility and bioreactivity (osteoconduction). The slow release system developed presents the advantage of first a complete release of the drug (few weeks) instead of some percentage with an acrylic bone cement matrix. It is degraded almost completely after a few months, whereas ceramic matrices take a very long time to degrade or are not degrade at all. The present study indicated that although the released drug is aggressive, the implants do not lead to local necrosis. Moreover, despite the high amounts of drug released, the circulating levels in the blood remained low and non-toxic. In the longer term, osteogenesis was even observed (Lebugle et al. 2002).

9.3 Synthesis of Drug Conjugates

The anti-tumor drug cisplatin was adsorbed over calcium phosphates of various sizes (nano and micro) and its cytotoxicity studied. Drug targeting to regional lymph node and drug delivery profiles were also checked to use it as an effective agent as an anti-metastases drug delivery system.

9.3.1 Nano-Calcium Phosphate

In most studies, nano-calcium phosphate (n-CaP) is synthesized by co-precipitation of equal volumes of a 30 mM $Ca(NO_3)_2$ (Sigma) solution and a 30 mM K_2HPO_4 (Sigma) solution both filtered through a 0.1 μm filtration device (Millipore, Boston, USA), followed by the immediate addition of 1.67(v/v)% of 0.2 μm filtered DARVAN®811 (sodium polyacrylate, $M_w = 3300$, R.T. Vanderbilt Company, Inc. Norwalk, CT, USA) as a dispersing agent. After 1 h of stirring, n-CaP pellets are collected by using a SORVALL RC5B Plus centrifuge at 12,000 rpm (20,076g) for 30 min. Before binding with cisplatin, this n-CaP pellet is re-dispersed in H_2O to wash off any ions lost from the surface which are then collected by centrifugation at 12,000 rpm for 30 min.

9.3.2 Apatitic Calcium Phosphate

Poorly crystalline hydroxyapatite crystals (denoted here as LTCA) are usually synthesized through the precipitation of two solutions. The first solution consists of 31.5 g of calcium nitrate and 0.45 g of magnesium chloride hexahydrate in 270 ml deionized distilled water. The second solution is prepared by combining 90 g of dibasic sodium phosphate, 36 g of sodium bicarbonate, and 0.45 g of decahydrate sodium pyrophosphate with 900 ml of deionized distilled water. The first solution is rapidly poured into the second and allowed to mature for 10 min. The formed precipitate is collected and rinsed thoroughly with 3 L of deionized distilled water by the Buchner funnel filtration method aided by a vacuum pump. Filter papers can vary but a good one is hardened ashless Whatman filter paper #540 with particle retention of 8 μm. The precipitate is placed on a lyophilizer (Virtis Freezemobile 12, Virtis Co., Gardiner, NY) to dry for 3–4 days. The crystals are then sieved to obtain a particle size of less than 125 μm. LTCA is also referred to as micro-CaP (m-CaP) throughout this chapter.

9.3.3 Aquated Cisplatin Preparation

Aquated Cisplatin is usually prepared by reacting a 90 mM $AgNO_3$ solution (Sigma) with a Cisplatin (CDDP, Sigma) solution (about 1000 $\mu g/ml$) at a 2:1 molar ratio. The reaction mixture is mixed on a thermal rocker (Lab-Line®, model 4637) for 12–24 h in the dark. The precipitate is removed by using a centrifuge (Beckman CPR) at 3000 rpm for 20 min. The supernatant is transferred to another tube and centrifuged two more times to thoroughly remove the AgCl precipitate. The remaining supernatant is filtered through a 0.2 μm filter. The final concentration of aquated cisplatin is determined by Pt analysis using atomic absorption spectrophotometry (AAS, model 5100, Perkin Elmer). This aquated cisplatin is diluted to a desired concentration before binding with calcium phosphate nanoparticles.

9.3.4 NanoCaP*CDDP Conjugate Synthesis

The water-washed and centrifuged n-CaP is bound with CDDP by following a standard conjugate preparation procedure. Briefly, 31.55 mg of a n-CaP pellet (which corresponds to 5 mg of dry CaP as determined by drying in an oven) is added to 0.625 ml of 20 mM Potassium Phosphate buffer (KPB, pH = 6), sonicated for 10 s, and added to 0.625 ml of aquated CDDP (Initial binding CDDP concentration C_0), and bound at 37°C at speed 5 on a LAB-LINE® themorocker (Model 4637, Barnstead Thermolyne, IL, USA) for 4 h. The bound conjugate above is centrifuged again at 12,000 rpm for 30 min. The supernatant which contains unbound CDDP is decanted and measured for the final binding supernatant CDDP concentration (C_f) determined by AAS. The pellet is washed with 0.25 mL 10 mM KPB buffer and centrifuged at 12,000 rpm, 30 min. The supernatant from this KPB wash is decanted and measured for CDDP concentration (C_{KPB}). The above KPB-washed pellet is added to 0.21 ml of a 0.9% NaCl solution (Baxter Healthcare Corporation), sonicated, and then put on the same rocker at 37°C at speed 5 for the burst release of some undesired CDDP for 30 min. The above solution is centrifuged again at 12,000 rpm for 30 min. The supernatant is decanted and measured for supernatant CDDP concentration (C_w) and the pellet collected is the n-CaP*CDDP conjugate. In the case that more conjugates are needed, the above protocol can still be followed by enlarging the volume of each solution used and mass of CaP used proportionally. The drug loading and loading efficiency of the nanoconjugates can be controlled by changing the initial aquated CDDP concentration (C_0). These conjugates are referred as nanoconjugates (n-conjugates) in this chapter.

9.3.5 MicroCaP*CDDP Conjugates Synthesis

A method that can be utilized for the binding process is a modification of the method used by Barroug et al. (2004). The CaP/CDDP conjugates are formulated by binding aquated cisplatin with various concentrations to the surface of the CaP particulates (LTCA) in a chloride-free buffer (KPB or NH_3PB). The use of phosphate containing buffer can prevent the dissolution of calcium phosphate crystals. A previous study has also demonstrated that chloride free phosphate buffer prefers the uptake of Cisplatin by HA crystals. For this, 100 mg of LTCA3 can be combined with 15 ml of cisplatin solution and 15 ml of chloride-free phosphate buffer. Protected from light, the samples can be placed on a Lab-Line model 4637 thermal rocker (Melrose Park, IL) at a speed setting of 5 (corresponding to 50 cycles per minute) and kept at 37°C for a 4 h binding time. After binding, samples can be centrifuged at 8000 rpm for 15 min, the supernatant extracted and preserved, and then the pellet washed with 6 ml of a 10 mM KPB solution. Resuspension of the pellet can occur by vortexing then a repeat centrifugation allowing for the removal of the wash. The sample can then be lyophilized to dryness for approximately 24 h. These conjugates are referred as microconjugates (m-conjugates) in this chapter.

9.4 Characterization of Nano- and Micro-conjugates

9.4.1 Transmission Electron Microscopy

The above formulated novel drug carriers have been characterized by numerous methods. For example, the morphology of the LTCA3 and n-conjugate particulates were characterized by transmission electron microscopy (TEM). For this, approximately 5 mg of LTCA3 powder and n-conjugates was dispersed in approximately 5 ml of ethanol and sonicated with an Ultrasonic 1000 L Cell Disruptor for 1 min then the supernatant liquid was immediately transferred by a micropipette to a 3 mm diameter Formvar coated copper TEM grid and dried. The samples on the TEM grid were analyzed using a 100cx JEOL TEM at 80 kV in brightfield (BF) modes. TEM images are presented in Fig. 9.1.

As seen in Fig. 9.1, the nanoparticles have a uniform size and are well dispersed, while LTCA (m-CaP) is aggregated but composed of nanoparticles as well thus resulting in larger particulate sizes. This observation has been confirmed with particle size measurements as discussed in the following section.

9.4.2 Particle Size and Z-potential Measurement

To verify TEM particle size and determine charge properties of the drug carrying nanoparticles, researchers have used particle sizers and zeta- (or Z-) potential measurements, respectively. For this, LTCA3, n-CaP, n-conjugates, and μ-conjugates were dispersed in ultrapure H_2O at about 1 mg/ml concentration by an ultrasonic 1000 l cell disruptor. The particle size and Z-potential of particulates was measured on a 90 plus particle sizer coupled with a Z-potential analyzer (Brookhaven Instruments, NY). Results from one study are shown in Table 9.1.

Table 9.1 shows that the n-CaP and nano-conjugates had a uniform size and were more negatively charged as compared to m-CaP and micro-conjugates. It was the

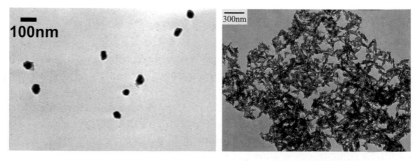

Fig. 9.1 TEM micrographs of n-conjugates and calcium phosphate (LTCA) particles before conjugation with cisplatin

Table 9.1 Particle size and Z-potential measurement for particulates and drug conjugates

	m-CaP	m-conjugates	n-CaP	n-conjugates
Particle size	7.5±6 μm	7.5±6 μm	117 nm	113.8 nm
Z-potential (mV)	−7.58	−11.23	−42.09	−51.23

motive of that study to obtain as much uniformity as possible for n-conjugates so that a size-mediated lymph node drug delivery carrier could be established. Also, pairs of n-CaP and n-conjugates, and m-CaP and m-conjugates were similar in size.

9.5 In Vitro Tests

9.5.1 Drug Loading

Of course, since these particles are intended to carry drugs, one must load drugs, perform release studies and in vitro analysis. For this, the adsorbed CDDP in nano- and micro-conjugates was calculated by the following equation:

$$\mu g \text{ Adsorbed CDDP} = (C_0^* V_0 - C_f^* V_f - C_{KPB}{}^* K_{KPB} - C_w^* V_w),$$

where V_0, V_{KPB}, and V_w are the volume of initial aquated CDDP, 10 mM KPB buffer, and NaCl used, respectively.

9.5.2 Drug Release

To determine drug release, 40 mg of nanoconjugates (88 μg/mg loading) and micro-conjugates (75 μg/mg loading) were dispersed in 0.8 ml PBS in an eppendorf tube. The conjugate was gently mixed with an 18 gauge needle and vortexed. It was then placed on a thermal rocker at 37°C at a rock speed of 2 (20 cycle/min) to release drugs at multiple time points. After 1 h, it was centrifuged at 9000 rpm for 10 min and the released drug from this 1 h release supernatant was measured by AAS. The pellet was added to 0.8 mL PBS again and mixed with a needle and put back on rocker at the same condition and released for 5 h. It was centrifuged again so the drug released at the 6 h time point was collected and measured. This procedure was repeated to obtain the drug release at 1, 3, 7, 12, and 16 days from both conjugates. Fig. 9.2 shows the results.

Results showed that micro-conjugates released more drugs and at a faster rate as compared to nano-conjugates (Fig. 9.2). This is due to the fact that there were higher electrostatic interactions between nano-particles and CDDP due to higher surface area and surface energy in the case of nanoparticles. Clearly this is desirable

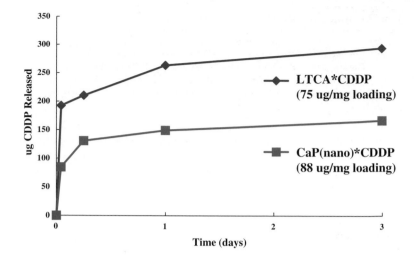

(a)

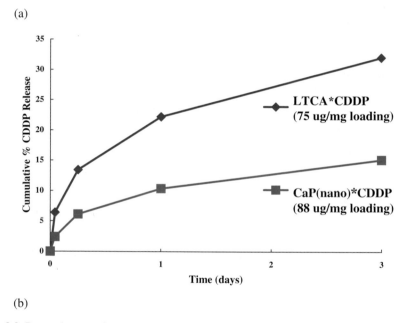

(b)

Fig. 9.2 Drug release profile of nano- and micro-conjugates. (**a**) Amount of CDDP released over time in PBS, PH 7.4. (**b**) Cumulative release over time of CDDP in PBS, PH 7.4

to keep the drug release and drug release rate lower for the case of nano-conjugates so that most of the CDDP can be delivered in the sentinel lymph nodes. This is one reason why so many researchers have been excited about the use of nanoparticles as drug release systems.

9.5.3 Cytotoxicity

9.5.3.1 Direct Addition of NanoCaP*CDDP Conjugates to A2780Cis Cells

However, a key question for any drug release system is if the delivery vehicle is toxic. This in general has been a concern for nanotechnology where many studies have not been conducted to date to determine their toxicity. As an initial attempt, 5 mg of nano- and micro-conjugates were synthesized in an all-sterile condition with a drug loading of 112 μg CDDP/mg CaP. Then 5 mg of these conjugate particles were dispersed in 0.8 mL PBS to give the highest concentration (if it is totally released, the CDDP concentration would be 700 μg/mL). Thus, conjugate suspensions were diluted 1:2 to 12 different concentrations. For a comparison, CDDP (control) was dissolved in saline around 1000 μg/mL and diluted in PBS at 200 μg/mL as the highest concentration. It was also diluted in 1:2 to 12 concentrations.

In order to test whether there was cytotoxicity from the n-CaP and LTCA3 alone, 5 mg of these particulates were dispersed in 0.8 mL of 200 μg/mL of the CDDP solution at the highest concentration. It was also diluted to a series of 12 concentrations. These concentrations were compared for cytotoxicity with a Free CDDP control.

Preliminary investigations of the growth rate of A2780cis human ovarian carcinoma cells were conducted in an effort to determine the proper concentration of cells to use for an in vitro cytotoxicity assays and direct cytotoxicity test of nanoconjugates. Using a Spectramax Plus384 spectrophotometer (Molecular Biosciences, Sunnyvale, CA), a direct correlation between cell concentration and observed absorbance values was made to determine if the examined cell concentration was within the linear range of detection. This linear range for A2780cis for each cell line was determined prior to the initiation of these studies. In order to obtain accurate results, the cell concentration of all samples, from those exposed to the highest amount of drug to the control cells receiving none, must be within the previously determined linear range at the end of the testing time period. Additionally, care must be taken that the cell concentration did exceed 80% confluency of the sample wells in order to prevent a change of cell growth rate that would potentially affect results.

After seeding the A2780cis cells in 96 well plates at 2000 cells/50μl/well for 24 h, 50 μl of Free CDDP (control) conjugate suspensions, n-Cap and LTCA3 with free drugs with 12 different concentration levels prepared as described earlier in the chapter were added to each well. For each sample, at least 5 replicates of every concentration were tested. The plates were then incubated for 44 h after drug addition, then 20 μl of CellTiter96® AQueous One (Promega) colorimetric proliferation reagent was added to each well, and then the plates were incubated for 4 more hours before being read on a Spectramax Plus384 spectrophotometer (Molecular Biosciences, Sunnyvale, CA) at an absorbance value of 490 nm. The celltiter 96® Aqueous assay has a tetrazolium compound. It is bioreduced by viable cells into a colored formanzan product, which is soluble in cell culture medium and has an absorbance maximum at 490 nm as related to the viable number of cells. Because particulates alone have interference at high concentrations around 490 nm, this absorbance will not reflect and be related to the number of viable cells. So this interference should be deducted for conjugates, n-CaP and LTCA3 in free drug. This interference was calibrated at the same condition as above (same seeding cell number, same conjugates and n-CaP and LTCA3 concentration and volume, same culture time) without the addition of celltiter assay.

9.5.3.2 Cytotoxicity Test of Released Drug from Nano-conjugates

In the experiments investigating the effects of drug release from n-conjugates on drug activity, 40 mg of n-conjugates at a loading of 35 μg/mg were dispersed in 0.8 ml PBS and put on rocker at 37°C at speed 2 (20 cycles/min) for 3 days. The 3-day released drug (supernatant after 9000 rpm centrifuge) was collected and the CDDP concentration was determined and used for a cytotoxicity test as compared to free CDDP which was prepared at the same time at a similar concentration. The procedure to test this released drug as compared to free drug was similar to the above.

9.5.3.3 IC50 Values

IC50 values (the concentration of drug that induces 50% inhibition of metabolic activity of the cells treated) are necessary for any new drug delivery vehicle, particularly for the numerous types of nanoparticles currently being researches. IC50 values have been determined on the above particles to assess the drug activity levels using the 4 parameter logistic equation:

$$Y = (A_{\max} - A_{\min})/1 + (x/\text{IC50})^n + A_{\min}$$

where Y = observed absorbance; $A_{\max}$ = absorbance of control cells; $A_{\min}$ = absorbance of cells in the presence of the highest agent concentration; x = drug concentration (μg/ml); and n = slope of curve.

Figure 9.3 shows how the interference from the particulate matter (n-CaP, m-CaP, n-conjugates, and m-conjugates) at 490 nm was subtracted to obtain the treated data which was then used to obtain IC50 values. IC50 values are shown in Fig. 9.4.

While the IC50 values for m-CaP and m-conjugates were similar or lower than controls CDDP, the n-CaP and n-conjugates were significantly different than controls. Moreover, in the case of nano-conjugates, the higher IC50 values showed that there will be reduced cytotoxicity at the site of injection. This is due to good electrostatic interactions between the n-CaP and CDDP. This drug will be made available to the tumor cells (which have an acidic microenvironment leading to the dissolution of CaP) and hence should be very effective for site-specific drug delivery. Also the drug released by the conjugates is as effective as the control itself. Although in this chapter, calcium phosphate-based nanoparticles are discussed, numerous studies have shown lower IC50 values for nanoparticle drug delivery systems.

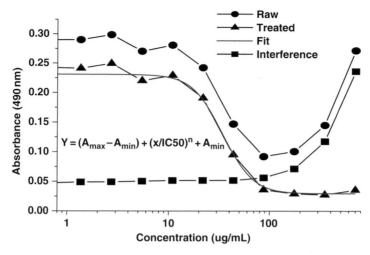

Fig. 9.3 Demonstration of IC50 value determination of nano-conjugates on A2780C in cancer cell lines. n-conjugate particles had interference around 490 nm at higher concentrations. This interference was calibrated at the same condition without adding the celltiter96 solution and was subtracted from raw data during cytotoxicity tests to give treated data (treated = raw-interference). Then 4-parameter sigmoidal fit of this treated data gave IC50 values

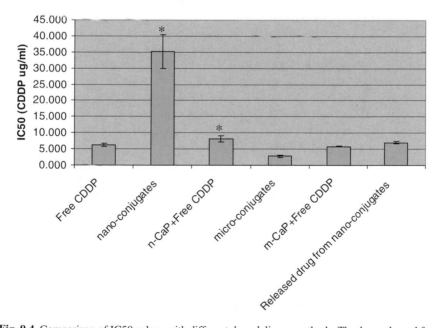

Fig. 9.4 Comparison of IC50 values with different drug delivery methods. The drug released from n-conjugates had a similar IC50 value as free CDDP indicating that the drug was viable and equally effective. * indicates p value <0.05 compared to all others

9.6 In vivo Tests

9.6.1 Drug Formulations

Although in vitro assays can provide valuable information for new cancer treatments and nanoparticle drug delivery strategies, in vivo assays are a must. Three sterile drug formulations have been prepared for such in vivo assays with the above-mentioned particles: (a) free drug solution, (b) micro-conjugates, and (c) nano-conjugates, specifically:

(a) CDDP drug solution of 1000 μg/ml in 0.9% saline,
(b) micro-conjugate formulations of concentrations of 50 mg/ml (dry weight of conjugates, determined by oven drying), and
(c) nano-conjugates formulation of 50 mg/ml.

Loading of drug formulations was determined by flame analysis on AAS and was determined to be 20, 60, and 90 μg of CDDP/mg of CaP. All the results shown here are normalized to compare for the same drug content.

9.6.2 Cytotoxicity

Of course, for any new nanoparticle material, cytotoxicity becomes important. As an example of such cytotoxicity studies, 12 female mice (Balb/c) were injected with 50 μl of PBS solution containing 1×10^5 66CL4 cells subcutaneously. The tumor was allowed to grow for 4 weeks in the mice following the injection. After that, n-conjugates (50 mg/ml CaP with loading of 102 μg of Cisplatin/mg of CaP) were injected intratumorally in the developed tumors at 50 and 100 μl (5 mice each), while 2 were kept as controls without any treatment. Figure 9.5 shows the change in tumor weight and volume with the novel nanoparticle treatment.

Weight change for the mice and tumor volume was measured at the 7th day of intratumoral injection. The tumor volume was calculated as (long axis)$^2 \times$ (short axis) $\times$ 0.4, where the long axis is tumor length around the major axis and short axis is the tumor length around the minor axis. Figure 9.5 showed that there was no toxicity effect as observed by the weight gain by the mice following the nano-conjugate injection (although the tumor growth was drug dosage dependent and was hampered more by the higher dosage (14.66 mg CDDP/kg of mice weight) compared to lower one (8.61 mg CDDP/kg of mice weight)). The control mice which were not treated died due to large tumor growth.

9.6.3 Evaluation of Drug Accumulation in Lymph Node

In addition to cytotoxicity, a drug needs to be evaluated for accumulation in the desirable tissue. This was assessed in the same study mentioned above. Specifically,

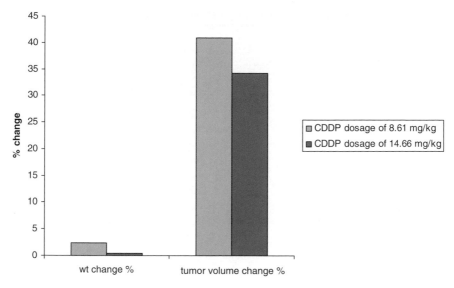

Fig. 9.5 Weight and tumor volume change after tumor induction and further in vivo drug conjugate injection

24 female Balb/c mice were maintained under recommended conditions. The mice (4 weeks old) were injected subcutaneously into the dormal surface of both footpads with free drug, micro-conjugate, and nano-conjugate formulations (20 μl each). Mice were euthanized using CO_2 asphyxiation at various time points of 30 min, 1 day, 4 day, and 7 day and both sides of the popliteal lymph node (sentinel node) were collected. The harvested lymph nodes and footpads were digested in 65% nitric acid for 2 days 37°C in a water bath. These homogeneous samples were then analyzed using graphite furnace analysis of AAS to get the CDDP amount in the sentinel lymph node. Figure 9.6 shows the results of drug accumulation in the sentinel node.

The transportation to lymph nodes depends largely on particle size and hence smaller ones enter first, while retention is also particle size dependent as larger particles localize in the lymph node while smaller ones pass on. The results of this present study support these findings.

Specifically, the free drug concentration was the highest at the 30 min period after it decreased to zero and no drug was observed at later time periods. There was a higher concentration of CDDP with m-conjugates compared to n-conjugates after 1 day. It was due to the fact that while as a whole m-conjugates were larger, they were composed of nanoparticles smaller in size than n-conjugates (Fig. 9.1). Some of these m-conjugate particles were loosely bound and with animal movements causing them to separate and hence higher concentrations of CDDP in the sentinel lymph node after Day 1, because initial lymph node transportation was particle size dependent.

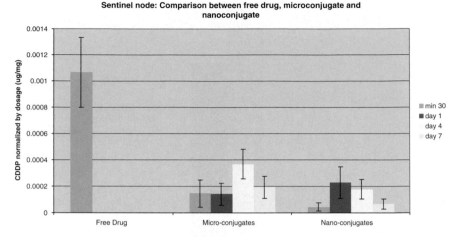

Fig. 9.6 The drug accumulation in sentinel node (popliteal node) following the drug treatment after various time intervals

At Day 1, the drug concentration with n-conjugates was higher compared to m-conjugates indicating that all the loose particles from m-conjugates were already in lymph nodes while n-conjugates were only then arriving.

Lastly, at Day 4 and Day 7 again the drug concentration was higher with m-conjugates as now the larger particles arrived at lymph nodes and they were retained while the smaller n-conjugates passed further on.

9.7 Conclusions

Nanomedicine has started to show promising results in tumor therapy. Nano- and micro-conjugates (cisplatin adsorbed on CaP) were emphasized here in their role in cancer treatment. The nano-conjugates were well dispersed and were uniform in size due to the addition of biocompatible surfactant DARVAN 811. Aquated CDDP was simply and efficiently adsorbed to the surface of CaP nanoparticles through electrostatic interactions. The high surface area of the CaP nanoparticles led to a reduction of the CDDP burst release and greater sustained release relative to more crystalline, micron-sized agglomerated particles of CaP. In vitro cytotoxicity testing showed that the CDDP released from the nanoconjugates retained complete activity during conjugation and release and had comparable cytotoxicity to free drug. The nanoCaP alone was not cytotoxic. The in vivo studies showed that there was no toxic effect of the drug formulations and higher dosage resulted in reduced tumor growth. Also the cisplatin was delivered to the regional lymph node and was dependent on conjugate size. This chapter thus highlighted that the favorable properties of nanoCaP/CDDP conjugates warrant their further investigation in intratumoral anti-cancer drug delivery applications. Anti-tumor drug CDDP bound nano- and micro-conjugates can be used for site-specific (lymph node) drug delivery.

Acknowledgments The author would like to thank the Susan G. Komen Foundation which supplied funds to carry out some of the work presented here.

References

Alitalo K, Tammela T, Petrova TV (2005) Lymphangiogenesis in development and human disease. Nature 438:946–953.

Barroug A, Kuhn LT, Gerstenfeld LC, Glimcher MJ (2004) Interactions of cisplatin with calcium phosphate nanoparticles: in vitro controlled adsorption and release. J Orthopedic Res 22: 703–708.

Bulte JWM (2005) Magnetic nanoparticles as markers for cellular MR imaging. *J Magn Magn Mater* 289:423–427.

Dadras SS, Paul T, Bertoncini J, Brown LF, Muzikansky A, Jackson DG, Ellwanger U, Garbe C, Mihm MC, Detmar M (2003) Tumor lymphangiogenesis: a novel prognostic indicator for cutaneous melanoma metastasis and survival. Am J Pathol 162:1951–1960.

Denekamp J. (1999) The tumour microcirculation as a target in cancer therapy: a clearer perspective. Eur J Clin Invest 29:733–736.

Fisher B, Fisher ER (1966) The interrelationship of hematogenous and lymphatic tumor cell dissemination. Surg, Gynecol Obstet 122:791–798.

Hall FT, Freeman JL, Asa SL, Jackson DG, Beasley NJ (2003) Intratumoral lymphatics and lymph node metastases in papillary thyroid carcinoma. Arch Otolaryngol-Head Neck Surg 129: 716–719.

Hashida M, Egawa M, Muranishi S, Sezaki, H (1997) Role of intra-muscular administration of water-in-oil emulsions as a method for increasing the delivery of anticancer agents to regional lymphatics. J Pharmacokinet Pharmacodyn 5:223–239.

Jussila L, Alitalo K (2002) Vascular growth factors and lymphangiogenesis. Physiol Rev 82: 673–700.

Kyzas PA, Geleff S, Batistatou A, Agnantis NJ, Stefanou D. (2005) Evidence for lymphangiogenesis and its prognostic implications in head and neck squamous cell carcinoma. J Pathol 206:170–177.

Lebugle A, Rodrigues A, Bonnevialle P, Voigt JJ, Canal P, Rodriguez F (2002) Study of implantable calcium phosphate systems for the slow release of methotrexate. Biomaterials 23:3517–3522.

Maula SM, Luukkaa M, Grenman R, Jackson D, Jalkanen S, Ristamaki R (2003) Intratumoral lymphatics are essential for the metastatic spread and prognosis in squamous cell carcinomas of the head and neck region. Cancer Res 63:1920–1926.

Muranishi S (1991) Drug targeting towards the lymphatics. *Adv Drug Res* 21:1–38.

Nakamoto Y, Fujiwara M, Noguchi T, Kimura T, Muranishi S, Sezaki H (1975) Studies on pharmaceutical modification of anticancer agent. I. Enhancement of lymphatic transport of mitomycin C by parenteral emulsions. Chem Pharm Bull 23:2232–2238.

Oussoren C, Storm G (1998) Targeting to lymph nodes by subcutaneous administration of liposomes. *Int J Pharm* 162:39–44.

Parker RJ, Priester ER, Sieber SM (1982) Comparison of lymphatic uptake, metabolism, excretion and biodistribution of free and liposome entrapped [14C]cytosine β-Image-arabinofuranoside following intraperitoneal administration to rats. Drug Metab Dispos 10:40–46

Pepper MS (2001) Lymphangiogenesis and tumor metastasis: myth or reality? Clin Cancer Res 7:462–468.

Reddy ST, Rehor A, Schmoekela HG, Hubbella JA, Swartz MA (2006) In vivo targeting of dendritic cells in lymph nodes with poly(propylene sulfide) nanoparticles. J Controlled Release 112:26–34.

Swartz MA (2001) The physiology of the lymphatic system. Adv Drug Deliv Rev 50:3–20.

Swartz MA, Skobe M (2001) Lymphatic function, lymphangiogenesis, and cancer metastasis. Microscopy Res Tech **55**:92–99.

Tahara Y, Ishii Y (2002) Apatite cement containing cis-diamminedichloroplatinum implanted in rabbit femur for sustained release of the anticancer drug and bone formation. J Orthop Sci 6:556–565.

Thiele W, Sleeman JP (2006) Tumor-induced lymphangiogenesis: a target for cancer therapy? J Biotechnol 124:224–241.

Weinstein JN (1983) Target-direction of liposomes: Four strategies for attacking tumor cells. In: Chabner B (ed) Rational basis of chemotherapy. A.R. Diss, New York, pp 441–473.

Chapter 10
Nanoparticles for Cancer Diagnosis and Therapy

Andrew Z. Wang, Frank X. Gu, and Omid C. Farokhzad

Abstract Nanotechnology enables unique approaches to the diagnosis and treatment of cancer. Over the last two decades, a large number of nanoparticle-based cancer diagnostics and therapeutics have been developed. These include superparamagnetic iron oxide nanoparticles, gold nanoparticles, liposomes and polymeric nanoparticles. While some of the liposomal and polymer-based therapeutic nanoparticles have advanced to clinical applications, a greater number of remaining nanoparticles platforms are currently in the preclinical stages of development. They have relatively incomplete toxicity profiles at this time. In this chapter, we will review the available preclinical and clinical toxicity data of a variety of nanoparticle platforms.

Contents

A.Z. Wang (✉)
Laboratory of Nanomedicine and Biomaterials, Department of Anesthesia, Brigham and Women's Hospital, Harvard Medical School
e-mail: zawang@partners.org

T.J. Webster (ed.), *Safety of Nanoparticles*, Nanostructure Science and Technology, DOI 10.1007/978-0-387-78608-7_10, © Springer Science+Business Media, LLC 2009

10.1 Introduction

Cancer is the second leading cause of death and accounts for 1 in 4 deaths in the USA, exceeded by only heart disease. An estimated 1,444,920 new cases of cancer will be diagnosed, and 559,560 Americans will die of cancer in 2007[1]. Once considered an incurable disease, modern medicine has made great strides in treating this complex and devastating illness. Today, most patients diagnosed with early stage cancer will survive their illness, as evidenced by the 5-year survival rate of 66% for all cancers diagnosed between 1996 and 2002 (up from 51% in 1975–1977). Advances in cancer diagnostics and therapeutics over the last several decades are largely responsible for this dramatic improvement. These advances include molecular imaging techniques, new chemotherapeutic agents and regimens, as well as biologically targeted therapies. To further improve cancer care, current efforts are focused on the development of novel cancer diagnostics and therapeutics.

Nanoparticles have been used as a modality of cancer therapy as early as the 1960s in the form of radiocolloidal gold [2]. Over the last two decades, a number of nanoscale particles – including metal/metal oxide nanoparticles, nanoshells, lipid-based nanoparticles, and polymer-based nanoparticles – have been under investigation. Nanoparticles can be engineered to encapsulate diagnostic and/or therapeutic agents, allowing them to serve as multifunctional and tumor-specific diagnostic/therapeutic agents. There are two main approaches for nanoparticles to target tumor cells: passive- and active-targeted therapeutics delivery. The passive targeting approach utilizes the unique properties of the tumor microenvironment: (1) leaky tumor vasculature which is highly permeable to macromolecules relative to normal tissue, and (2) a dysfunctional lymphatic drainage system which results in enhanced fluid retention in the tumor interstitial space. As a result of these characteristics, the concentration of polymeric NPs and macromolecular assemblies found in tumor tissues can reach up to 100 times higher than those in normal tissues [3]. The biggest limitation of passive tumor targeting is the inability to achieve a sufficiently high level of drug concentration at the tumor site, resulting in low therapeutic efficacy and eliciting undesirable systemic adverse effects [4]. In comparison with passive targeting which utilizes pharmacokinetics manipulation and NP size reduction to achieve EPR, active targeting is achieved by delivering drug-encapsulated NPs to uniquely identified sites while having minimal undesired effects elsewhere. Active tumor targeting is typically achieved by both local and systemic administration of NPs with targeting molecules conjugated on the particle surface; these targeting molecules recognize and bind to specific ligands that are unique to cancer cells. The cytotoxic drug encapsulated in the NPs can thus be delivered directly to cancer cells while minimizing harmful toxicity to non-cancerous cells adjacent to the targeted tissue. Both passive- and active-targeted nanoparticles have been developed for each of the nanoparticle platforms; we will review these various nanoparticles for cancer applications and their safety/toxicity in this chapter.

10.2 Superparamagnetic Iron Oxide Nanoparticles

10.2.1 Background

Small iron oxide particles have been under investigation as diagnostic agents for nearly four decades [5]. In the last decade, the focus has been on nanoscale iron oxide particles, such as maghemite (γFe_2O_3), magnetite (Fe_3O_4), and other ferrites. Iron oxide nanoparticles are unique, since they possess characteristics that neither the atom nor larger particles have [6]. These iron oxide nanoparticles exhibit superparamagnetic phenomena, characterized by a large magnetic moment in the presence of a static external magnetic field, making them excellent contrast agents for magnetic resonance imaging (MRI) [7].

10.2.2 Formulation

Conventionally, iron oxide nanoparticles (either Fe_2O_3 or Fe_3O_4) are synthesized by co-precipitation of Fe^{2+} and Fe^{3+} aqueous salts with the addition of a base [8]. The size, shape, and composition of iron oxide crystals depend on the type of salts used, Fe^{2+} to Fe^{3+} iron ratio, pH, and ionic strength of the media [8]. After synthesis of the magnetic nanoparticles, surface modification is required to make the particles soluble. Surface modifications also prevent the destabilization and aggregation of the particles. Iron oxide nanoparticles can be coated by organic non-polymeric molecules, organic polymers, and inorganic molecules. Biocompatible polymers and monomers are the most common surface coating molecules for biomedical applications. These biocompatible materials include dextran, polyethylene glycol (PEG), polyvinyl alcohol (PVA), amino acids, and citric acid. The nanoparticle surface can be further functionalized with biological molecules such as antibodies, proteins, and targeting ligands. These biological molecules are conjugated to polymer surfaces via amide or ester bonds to make the iron oxide nanoparticles target-specific.

10.2.3 Applications for Iron Oxide Nanoparticles

The most important application for superparamagnetic iron oxide nanoparticles (SPIO) is in MRI. Currently, there are two main classes of SPIO in clinical use: SPIO with a mean diameter greater than 50 nm, and ultrasmall superparamagnetic iron oxide nanoparticles (USPIO) with diameter less than 50 nm [9]. The larger SPIOs, which include ferumoxides (Feridex) and ferucarbotran (Resovist), are used for liver imaging. Because of their larger size and the dextran coating, they are preferentially taken up by Kupffer cells (macrophages) in the liver and produce a dark signal on T2 sequence on MRI. In this way, liver tumors can be identified by their lack of Kupffer cells, appearing bright on T2 sequence. USPIOs, such as ferumoxtran-10 (Combidex), have longer circulation time and are useful in identifying metastatic cancer in the lymph nodes. In a Phase III clinical trial, ferumoxtran-10 has been shown to have 90.5% sensitivity and 97.8% specificity for prostate cancer nodal disease [10].

In addition to their application as imaging agents, SPIO have been shown to be potential drug delivery vehicles and hyperthermia agents [11,12]. For example, Kohler et al. formulated a biostable methotrexate (MTX)-immobilized iron oxide nanoparticle for the delivery of methotrexate [13]. SPIO, under alternating magnetic fields releases heat and can be utilized as hyperthermia agents. Wust et al. demonstrated the potential of iron oxide nanoparticles as hyperthermia agents in a study of 22 patients where a temperature of 40°C was achieved in tumors in most of the patients [14].

10.2.4 Toxicity

Iron oxide nanoparticles are usually taken up by macrophages in the mononuclear phagocytic system (MPS) of the liver, spleen, lymphatics, and bone marrow [15]. The blood half-lives of iron oxide nanoparticles vary from 1 h (VSOP-C184) to 24–36 h (ferumoxtran-10) [16,17]. Their half-life is dependent on surface coating material as well as particle size. Given the same particle size, the half-life of ionic dextran-coated nanoparticle ferumoxytol is shorter than that of non-ionic dextran-coated ferumoxtran-10 (10–14 h vs. 24–36 h) [18]. Given the same surface coating material, smaller iron oxide nanoparticles have longer half-lives than do the larger ones. For example, ferumoxtran-10 (30 nm) has a half-life of 24–36 h whereas ferumoxides (120–180 nm) has a half-life of 2 h. Nanoparticles with longer circulation times will have greater uptake into MPS cells of organs other than liver, whereas sinusoidal liver cells will take up the majority of nanoparticles with shorter blood half-lives [19].

Once taken up by MPS cells, the iron oxide nanoparticles are broken down through the lysosomal pathway [20]. Intracellular metabolism is dependent on nanoparticle composition. Schultz et al. studied the degradation of ferumoxtran using ^{14}C and ^{59}Fe double-labeled nanoparticles[20]. The dextran coating was degraded by intracellular dextranases and excreted in the urine (89% in 56 days). The degraded iron was incorporated into the body's iron storage and was found in red blood cells in the form of hemoglobin. Like endogenous iron, it is eliminated very slowly, as only 16–21% of the iron injected was shown to be eliminated after 84 days. Other studies showed that degraded iron can be stored in the forms of ferritin and/or transferrin [11,17,21]. Briley-Saebo et al. showed that a single injection of feruglose (5 mg/kg) increases the iron content of the liver 3 days post-injection, supporting the hypothesis of metabolized iron stored as ferritin and transferrin [22]. Phase II clinical trial of feruglose also showed that the liver remains hypointense on T1-weighted images for several months after a single injection [23].

The toxicity of iron and its derivatives in vivo has been well studied. Normal liver contains approximately 0.2 mg of iron per gram, and total human iron stores amount to 3500 mg. The amount of iron in iron oxide imaging varies between 50 and 200 mg, which is small when compared to the total body storage. Since chronic iron toxicity develops only when the iron concentration in liver exceeds 4 mg Fe/g, iron oxide nanoparticles are unlikely to cause any long-term toxicity due to the iron load [24].

Another source of potential toxicity is the surface coating molecules in iron oxide nanoparticles. The presence of anti-dextran antibodies in individuals after USPIO administration suggests such risk [25]. Today, most of the iron oxide nanoparticles under clinical investigation are coated with dextran, and no anaphylactoid reactions have been reported. Such reactions to dextrans 40 and 70 have been well documented in other clinical scenerios [26]. One potential explanation is that dextran used for iron oxide nanoparticles are much smaller (2–10 kDa). Citrate-coated iron oxide nanoparticles have also been shown to cause significant increase in oxidative stress when taken up by macrophages. But this oxidative stress is only transient and do not appear to affect cellular proliferation [27]. Most iron oxide nanoparticles do not seem to trigger cell activation, given their good biocompatibility profiles. For example, in a study evaluating ferumoxtran-10, interleukin-1 was not released during in vitro endocytosis by macrophages [28].

Despite the increasing interest in the delivery of magnetic nanoparticles into various cells types, their toxicity on cell types other than the MPS cells has not been well studied. Pisanic et al. studied the nanotoxicity of iron oxide nanoparticle internalization in growing neurons [29]. Maghemite nanoparticles coated by dimer-captosuccinic acid (DMSA) were delivered into rat pheochromocytoma cell line PC12M. Statistically significant reduction in PC12 cell viability and cell detachment were observed within 48 h of exposure to the nanoparticles. They also showed that with increasing concentrations of iron oxide nanoparticles, PC12 cells demonstrated dose-dependent diminishing viability and capacity to extend neuritis in response to nerve growth factors. These results highlight the importance of further study into the acute and long-term effects of iron oxide nanoparticles in cell types outside of the MPS.

Preclinical evaluations of SPIO and USPIO showed satisfactory profiles according to standard toxicological and pharmacological tests. The first SPIO agent used in clinical trials was ferumoxides (Feridex), which has a mean particle diameter of 160 nm and a half-life of 8 min. In a Phase I trial with 20 patients with liver metastases, no serious clinical side effects occurred [28]. One patient reported a sensation of anxiety which was of no clinical significance. There was no significant change in the patients' heart rate, blood pressure, or urine analytes. Laboratory parameters that changed with administration of Feridex included protein level, serum iron level, transferrin, and ferritin levels. Another trial using Feridex involved 208 patients; no patient had serious adverse reaction to the ferumoxides [30]. Most symptoms were mild to moderate excepting two patients who experienced severe back pain and one patient who had severe flushing. None of the side effects were of clinical significance. There was a statistically significant increase in mean percentage saturation of total iron-binding protein, serum iron concentration, and ferritin level. Another SPIO, SH U 555A (Resovist), was also found to be safe and effective in clinical trials, with a similar toxicity profile to Feridex [31]. The USPIO ferumoxtran (Combidex) is still pending FDA approval. In a Phase I trial involving 41 healthy patients, it had no clinically significant post-dose changes in physical examination findings, vital signs, or electrocardiogram results. No significant changes in laboratory values were noted by the investigators. All fourteen adverse events reported were considered not serious and not clinically significant. In a Phase II study of 104 patients with liver or spleen lesions who underwent ferumoxtran imaging, 15% of the patients reported a total of 33 adverse events [32]. The side effects included dyspnea (3.8%), chest pain (2.9%), and rash (2.9%). No serious adverse events were reported during the 48-hr observation periods and there were no clinically significant effects. Harisinghani et al. used ferumoxtran for the detection of prostate cancer lymph node metastases [10]. In a study with 80 patients, only 5 patients reported back pain during contrast infusion, and reported episodes of pain resolved without necessitating intervention.

Several clinical studies from Germany evaluated the feasibility of using iron oxide nanoparticles as hyperthermia agents. Johannsen et al. carried out a pilot study using magnetic nanoparticles for interstitial hyperthermia in recurrent prostate cancer [33]. The patient tolerated the treatment well without anesthesia and without side effects. Maier-Hauff et al. used iron oxide nanoparticles for hyperthermia in 22 patients with non-resected or recurrent glioblastoma multiforme [34]. Iron oxide nanoparticles were injected into the tumor bed under anesthesia. Thermotherapy was well tolerated, and only minor side effects were observed. Symptoms included a temporary rise in blood pressure and transient deterioration of neurological symptoms.

10.3 Gold Nanoparticles

10.3.1 Background

Gold particles have been used as therapeutic agents for the treatment of human diseases such as rheumatoid arthritis for many decades [35,36]. Gold nanoparticles are particularly attractive for use in biological applications for several reasons. First, gold is a noble metal with inert chemical properties, resistant to corrosion, and has low toxicity based on past clinical experience. Gold nanoparticles are also relatively easy to synthesize [37]. Lastly, the gold surface can be easily functionalized with biological molecules, such as antibodies and nucleic acids. There are two types of nanosized gold particles for biomedical use: gold colloids and gold nanoshells. Gold nanoshells are a unique class of nanoparticles because they have tunable plasmon resonance, and, similar to quantum dots (QDs), can be used as in vivo imaging agents.

10.3.2 Formulation

Gold colloids are generally produced in a liquid by reducing hydrogen tetra-chloroaurate ($HAuCl_4$). Reduction of Au^{3+} ions to unionized gold atom leads to a supersaturated gold solution. The gold gradually precipitates in the form of nanometer-sized particles, leading to gold colloids.

Gold nanoshells consist of a spherical dielectric nanoparticle core surrounded by a thin gold shell. The first nanoshell was developed by Zhou et al. and consisted of an Au_2S core covered by a gold shell [38]. The Au_2S–Au nanoparticles were grown in a one-step process, where hydrogen tetrachloroaurate and sodium sulfide are mixed. This synthesis cannot control core and shell thickness. Oldenburg et al. then developed a new particle, silica core gold nanoshells [39]. The dielectric silica core was first grown, followed by functionalization of the surface with amine groups, and lastly, small gold colloids (1–2 nm) were absorbed onto the silica surface.

10.3.3 Applications

Gold nanoparticles have been utilized as both imaging and therapeutic agents for cancer. Sokolov et al. used gold nanoparticles conjugated to EGFR antibodies as cell imaging agents to label cervical biopsies for the identification of precancerous lesions [40]. Gold colloid nanoparticles were also used as drug delivery vehicles. Paciotti et al. showed that TNF-conjugated gold nanoparticles can prolong survival in a murine model of colon cancer [37]. Another application for gold nanoparticles is enhancement of tumor cell sensitivity to external beam radiotherapy. Hainfeld et al. showed that systematically injected gold nanoparticles accumulated in a murine subcutaneous tumor model and enhanced the effect of radiotherapy [41].

Gold nanoshells have been used as in vivo contrast agents for imaging with optical coherence tomography, diffuse optic tomography, and photoacoustic tomography [42,43]. They have also been utilized as nanoscale drug carriers. The ability of nanoshells to act as photoabsorbers and to release heat made them potential photothermal ablation agents. Several in vivo animal studies have been conducted using nanoshells as photothermal ablation agents [43–45].

10.3.4 Toxicity

Gold has three states: elemental gold, gold I, and gold III. The pharmacological and toxicological profiles of the three states are very different [46]. Elemental gold is chemically one of the least active metals. It does not react to strong alkalis or acids with the exception of selenic acid and aqua regia [46]. Metallic gold is generally non-toxic and is used regularly in dental implants as well as in food decorations. Side effects include skin reactions and rare instances of lichen planus [47].

Gold (I) salts have been used for medical treatments as early as 1890, when Robert Koch showed the cytotoxic effect of gold (I) salts on the tubercle bacillus [48]. In the 1930s, gold (I) salts became one of the treatments for rheumatoid arthritis. After more than 60 years of use, gold (I) remains one of the standard medications for this disease with a dozen parenteral and oral gold formulations on the market.

After absorption, gold salts, mostly bound to albumin and globulin, distribute throughout the body [49,50]. Gold salts can remain in the plasma for many months, but most of the gold will ultimately become fixed in the tissues, mainly in the kidneys, liver, spleen, marrow, skin, hair, and nails [50,51]. Upon intracellular uptake, gold salts are contained in phagolysosomes called aurosomes [51,52]. The rest of the gold salts are excreted in the urine [53]. The metabolism of metal gold can also lead to gold (I). Elemental gold can be dissolved at minute levels by thiol-containing molecules such as cysteine, penicillamine, and glutathione to yield gold (I) complexes [54]. Brown et al. simulated conditions inside the phagocytic lysosomes and showed substantial dissolution of gold (0) in the presence of hydrogen peroxide and amino acids [55]. There is a large amount of clinical data on the clinical use of gold (I) salts from arthritis treatments. A Cochrane review of injectable gold for rheumatoid arthritis looked at four trials and 415 patients [56].The toxicity analysis showed that 22% of treated patients withdrew from treatment compared to 4% of the controls. Most of the patients withdrew due to skin reactions, and only 3% of the patients discontinued due to leukopenia and thrombocytopenia. In general, dermatitis represents two-thirds of all adverse reactions to chrysotherapy (gold therapy), with approximately half of treated patients experiencing one form or another of dermatitis [57]. Common skin toxicities include pruritus, rash, cheilitis, chronic popular eruptions, contact sensitivity, erythema nodosum, allergic contact purpura, pityriasis rosea, and lichenoid and exfoliative dermatitis. Non-dermatologic and more serious reactions include blood dyscrasias, eosinophilia pulmonitis, nephrotoxicity, and nephrotic syndrome [48,58].

Gold (III) is the most reactive and can cause oxidation of methionine residues to sulfoxides, leading to protein denaturation [59]. Schuhmann et al. demonstrated using murine models that gold (III) can elicit strong immune responses, and these dose-dependent responses were both T-cell dependent and specific [60]. Reactive gold (III) species become reduced to gold (I) after oxidation reactions and subsequently follow gold (I) metabolization.

Although there is no clinical data on the use of gold nanoparticles for cancer diagnosis and treatment, there are a number of in vitro and in vivo studies on the toxicity of gold nanoparticles for cancer. Connor et al. tested the toxicity of gold spheres of 4, 12, and 18 nm using the K562 leukemia cell line [61]. They demonstrated that gold nanoparticles are not inherently toxic to human cells based on the MTT study. Goodman et al. compared the toxicity of gold nanoparticles functionalized with cationic and anionic side chains [62]. They found that while cationic particles are moderately toxic based on MTT, hemolysis, and bacterial viability assays, anionic particles are quite non-toxic. Intracellular uptake of gold nanoparticles has also been shown to be non-toxic to macrophages and does not elicit proinflammatory cytokines [63].

Animal in vivo studies have also been conducted using gold nanoparticles. Hainfeld et al. injected 2 nm sized gold nanoparticles into mice as radiotherapy sensitizers [41]. Pharmacokinetics showed an early rapid rise followed by a slower clearance rate. At 5 min following injection, the tumor to muscle gold ratio was 3.5:1.0. Mice received 2.7 g Au/Kg of gold and lived for more than one year without overt clinical signs. O'Neal et al. injected gold nanoshells intratumorally in mice as photothermal agents [45]. At 90 days post injection, the mice were healthy and had no observable side effects. These in vivo studies support the in vitro toxicity profile of gold nanoparticles.

10.4 Quantum Dots

10.4.1 Background

Quantum dots (QDs) are nanometer-sized (1–12 nm) semiconductor nanocrystals with unique optical and electrical properties [64,65]. Characteristics of QDs include broad band excitation, narrow bandwidth emission, emission of high intensity light, resistance to quenching, and excellent photochemical stability. These properties lead to the rapid development of QDs as potential new fluorescent probes for biomedical imaging.

10.4.2 Formulation

The QDs for biological applications generally consist of a metalloid crystalline core and a biocompatible surface. QD cores are formulated from a variety of metal complexes, such as semiconductors, noble metals, and magnetic transition metals. Common cores include zinc sulfides (ZnS), zinc-selenium (ZnSe), cadmium-selenium (CdSe), cadmium-tellurium (CdTe), indium phosphate (InP), indium arsenate (InAs), gallium arsenate (GaAs), and gallium nitride (GaN) [65,66]. Heavier QDs such as CdTe/CdSe and CdSe/ZnTe have also been established [67].

QD cores are inherently hydrophobic and bio-incompatible. In order to make them biologically useful, they need to be surface functionalized with biocompatible molecules. Surface coating allows (i) the solubilization and stabilization of QDs in biological buffers, (ii) maintenance of their original colloidal and photophysical properties, and (iii) reactive group availability for subsequent conjugation to bio-molecules. Molecules that have been used for QD surface modifications include thiol-containing molecules, oligomeric phosphines, dendrons, peptides, diblock or triblock copolymers, silica shells, phospholipids micelles, polymer shells, and polysaccharides [64,67,68]. The QD surface can be further functionalized by targeting molecules such as antibodies, peptides, aptamers, etc [69].

10.4.3 Applications

The unique photochemical properties of QDs lead to their rapid development as in vivo fluorescent imaging probes. QDs were shown to have high-sensitivity and high-contrast imaging in deep tissues in mice as well as in larger species [70]. The first tumor targeting application of QDs was demonstrated by Gao et al. using PEG and antibody-coated QDs [71]. Kim et al. used CdTe/CdSe QDs to perform nearly background-free imaging of lymph nodes 1 cm deep in tissues [70]. QDs have yet to be used as an imaging agent in the clinical setting despite their vast potential, due to concerns regarding their toxicity.

QDs have also been shown to be potential photodynamic therapy (PDT) agents. Samia et al. demonstrated CdSe QDs can be used to sensitize either a PDT agent or to act as a PDT agent alone by creating oxygen free radicals [72]. Bakalova et al. also showed that CdSe QDs conjugated to anti-CD antibody can sensitize leukemia cells to UV radiation and can potentiate the effect of photosensitizers [73].

10.4.4 Toxicity

QD toxicity depends on multiple factors, including particle composition, particle size, concentration, surface coating material, oxidative and photolytic properties [74]. Cadmium and selenium, two of the most widely used metals in QD cores, can cause both acute and chronic toxicities. For example, cadmium has a half-life of 20 years and is a suspected carcinogen that biodistributes in all tissues with no known mechanisms of excretion [75]. Several studies have looked at the toxicity of Cd-based QDs using extracellular QD concentrations. Derfus et al. and Kirchner et al. both demonstrated that water-soluble CdSe and CdSe/ZnS QDs can cause direct extracellular cytotoxicity [76]. The mechanism of cytotoxicity was due to $Cd2+$ ions released from QD photo-oxidation and degradation and can be minimized by surface coating of the QDs. For example, QDs coated with simple molecules, such as mercaptoacetic acid, mercaptopropionic acid, 11-mercaptoundecanoic acid, are more toxic than those coated with silica, which can better prevent Cd^{2+} from leaking [76,77]. Hoshino et al. also showed that QDs toxicity is highly dependent on the surface molecules by using comet assay, flow cytometry and MTT assay [77]. ZnS-coated CdSe QDs coated with MUA, cysteamine, and thioglycerol were used for the study. By comparing the non-coated QDs and the coated QDs, they concluded that the type of surface modification molecules is the most important factor in determining cytotoxicity.

The effect of size on QD toxicity was studied by Lovric et al. They found that cytotoxicity was more pronounced with smaller positively charged QDs (2.2 ± 0.1 nm) than with larger equally charged QDs (5.2 $\pm$ 0.1 nm) at equal concentrations [78]. Furthermore, the intracellular distribution was different for the different size particles, with the smaller cationic QDs localized to the nuclear compartment whereas the larger QDs localized to the cytosol.

Chang et al. demonstrated that cytotoxicity is also dependent on intracellular level of nanoparticles [79]. Bare CdSe/CdS-based QDs and CdSe/CdS-based QDs with different PEG surfaces at various concentrations were evaluated. They showed that given the same intracellular QD concentration per cell, the cytotoxicities were similar despite different surface modifications. In a different study, Chang et al. demonstrated that surface coating affects QD uptake and in turn influenced the intracellular cytotoxicity [80]. Choi et al. explored the mechanisms of intracellular toxicity using neuroblastoma cells [81]. CdTe QDs were incubated with human neuroblastoma (SH-SY5Y) cells. Cytotoxicity was correlated with Fas upregulation on the surface of the QD-treated cells. In addition, these cells had increased membrane lipid peroxidation. The peroxidized lipids were detected at the mitochondrial level, leading to impaired mitochondrial function. Chan et al. conducted a similar study on the mechanisms of QD-induced apoptosis [82]. Using CdSe QDs and neuroblastoma cell line IMR-32, they found that CdSe QDs can induce apoptotic pathways, including JNK activation, loss of mitochondrial membrane potential, mitochondrial release of cytochrome c, and activation of caspases 3 and 9. The QDs also triggered an increase in reactive oxygen species and inhibited survival signaling events such as RAS and Raf-1 signaling. It is also important to note that the same mechanisms for making QDs potential photosensitizers can also cause cellular toxicity. QDs can create free oxygen radicals which are highly cytotoxic.

Although there has been no human clinical trials using QDs, a number of in vivo animals studies have been conducted [71,83]. Akerman et al. injected targeting peptide-coated and ZnS-capped CdSe QDs into the tail veins of mice. They found that the injected QDs accumulated in the liver and spleen regardless of coating. There was no observable effect from the QD injection. Dubertret et al. encapsulated QDs in phospholipids block-copolymer micelles and injected them into *Xenopus* embryos. The QD-micelles were stable, nontoxic ($<5 \times 10^9$ QD/cell) and did not affect embryogenesis. Jaiswal et al. studied the effect of QDs on cell growth and cell development. They found that DHLA-capped QDs had no effect on Hela cells and no effect on the development of *D. discoideum* cells. In another in vivo study, Voura et al. injected QD-labeled tumor cells into mice. QDs had no detectable toxicity to the labeled cells or the host animal. Both Ballou et al. and Gao et al. conducted in vivo imaging of tumor in murine models. Ballou et al. showed that circulating half-lives of QDs are dependent on the surface coating (<12 min for amp, m-PEG 750 modified QDs vs. 50–100 min for the COOH-PEG-3400 QDs) and neither study showed toxicity by the administration of QDs.

Given the somewhat conflicting toxicity data between in vitro and in vivo studies, more in vivo studies with longer observation periods are needed to fully understand the toxicity effects of QDs on living systems.

10.5 Dendrimers

10.5.1 Background

Dendrimers are well-defined, regularly branched macromolecules with sizes in the 2.5–10 nm range [84]. Advantages of dendrimers include nanoscale spherical architecture, narrow polydispersity, multifunctional surface, and large surface area. Their relatively empty intramolecular cavity can be utilized to carry drugs, while their surface can be engineered to provide precise spacing of surface molecules as well as conjugation to targeting molecules. Since their first description by Vögtle in 1978, there has been an explosion of interest in dendrimers and their biomedical applications in recent years.

10.5.2 Formulation

Dendrimers are synthesized using either synthetic or natural elements such as amino acids, sugars, and nucleotides as the basic building blocks [84]. Their surface is frequently further modified with other biocompatible molecules. There are two major strategies for dendrimer synthesis. One uses the "divergent method" in which growth of a dendron originates from a core site, and the monomeric modules are assembled in a radial, branch-upon-branch fashion [85]. The other method follows a "convergent growth process;" this synthesis process starts from what will be the dendrimer surface and grows inward to a reactive focal point [86]. Using these strategies, over 100 dendrimer families and over 1000 different surface modifications have been reported [87]. Among them, the polyamidoamine (PAMAM) and poly(propylenemine) (PPI) have been most widely used for biomedical applications.

10.5.3 Applications

Potential applications of dendrimers in cancer include development as drug-carriers for targeted and controlled release of therapeutics, as vectors for oligonucleotide and gene delivery, and as diagnostic imaging agents. Malik et al. demonstrated the feasibility of dendrimers as carriers of chemotherapy by encapsulating cisplatin within

PANAM dendrimers [88]. They showed that the conjugates had slower drug release, higher accumulation in solid tumors, and lower toxicity when compared to free cisplatin. Boronated dendrimers are under investigation as agents for boron neutron capture therapy (BNCT) [89–91]. Dendrimers have also been studied as non-viral gene transfer agents and as delivery vehicles for antisense nucleotides and siRNA [92]. Gadolinium chelates conjugated to dendrimers have been evaluated as potential MRI contrast agents, especially for tumor imaging [93].

10.5.4 Toxicity

Many studies have examined dendrimer toxicity in vitro, but they have been conducted in different cell lines and under different experimental conditions. General trends show that dendrimers bearing $-NH_2$ termini display concentration and generation-dependent cytotoxicity. For example, Roberts et al. showed that cationic PAMAM dendrimers caused a decrease in cell viability with increasing concentrations in V79 Chinese hamster lung fibroblasts [94]. The concentration producing 90% cell death was 1 nM for generation 3, 10 nM for generation 5 and 100 nM for generation 7. One mechanism for dendrimer cytotoxicity is through membrane disruption. Hong et al. showed that generation 7 PAMAM dendrimers caused formation of holes in lipid bilayers, whereas generation 5 dendrimers were able to expand existing membrane defects [95].

Although dendrimer cytotoxicity is dependent on the chemistry of the core, it is most strongly influenced by the dendrimer surface. Chen et al. showed that cationic dendrimers were much more cytotoxic than anionic or PEGylated dendrimers using the MTT assay [96]. PAMAM-OH derivatives showed a lower level of cytotoxicity than PAMAM-NH_2 due to shielding of the internal cationic charges by surface hydroxyl groups [97]. The processes of increased branching (generation) and surface coverage with biocompatible groups such as PEG have been used to minimize surface toxicity.

Cationic dendrimers such as PAMAM, PPI, and diaminoethane (DAE) have been shown to cause red blood cell hemolysis above a concentration of 1 mg/ml [98]. Even at a non-hemolytic concentration of 10 μg/ml, PANAM and PPI dendrimers can cause changes in red blood cell morphology after 1 h of incubation. Plank et al. also demonstrated that PANAM dendrimers and high molecular weight poly-L-lysine (PLL) and polyethyleneimine (PEI) can cause complement activation [99].

The toxicity of dendrimers has also been evaluated using gastrointestinal tract models. Jevprasesphant et al. showed that generation 4 PAMAM dendrimers are toxic in a concentration-dependent fashion, causing barrier breakdown at 100 μg/ml using CaCo-2 cells [100]. They also demonstrated that while anionic PAMAM dendrimers had no effect on the transepithelial resistance (TEER), cationic PAMAM dendrimers reduced the TEER markedly.

The biodistribution of dendrimers through parenteral administration has been studied with the development of dendrimer-based MRI contrast agents and tumor targeting therapeutic agents. Malik et al. showed that ^{125}I-labelled cationic (generations 3 and 4) and anionic (generations 2.5, 3.5, and 5.5) PAMAM dendrimers were rapidly cleared from the circulation in Wistar rats [98]. When compared to cationic dendrimers, anionic dendrimers showed longer circulation times. Both showed generation-dependent clearance rates. When administered intraperitoneally, ^{125}I-labelled PAMAM dendrimers were transferred into the bloodstream within 1 h.

Dendrimers have been conjugated to a number of targeting ligands as targeted drug delivery systems, including folate, epidermal growth factor (EGF), and monoclonal antibodies [91,101]. However, in vivo studies have not demonstrated effective tumor targeting. When ^{131}I-labelled boronated-PAMAM-EGF conjugate was injected intravenously into rats bearing a C6-EGF transfected glioma, only 0.01% (24 h) and 0.006% dose/g (48 h) tumor localization was achieved [102]. Nonspecific liver and spleen uptake was much higher (5–12% dose/g). Tumor localization was much better using intratumoral injection; however it is not a practical form to administer treatment [89].

Few studies on dendrimer toxicity in vivo have been reported. Roberts et al. administered generations 3, 5, and 7 cationic PAMAM dendrimer to mice in either a single dose or repeated dose once per week for 10 weeks [94]. Three animals died in the generation 7 group and no other adverse effects (behavior changes or weight loss) were observed in the other groups. A number of biodistribution studies with dendrimer-based contrast agents have been reported. It has been shown that generation 4 PAMAM-based contrast agent is excreted in 2 days [103]. The contrast agent transiently accumulate in the renal tubules, lead to potential toxicity from Gd (III) ions released from the Gd (III) chelate. Kobayashi et al. showed that PPI-based imaging agents had higher liver accumulation when compared to the same generation of PAMAM-based agents [104].

A human clinical study has been conducted using dendrimer-based MRI contrast agent, SH L 643A (Gadomer-17). It was given to 12 volunteers for coronary artery imaging [105]. Compared with baseline values obtained prior to the examination, no significant changes in pulse rate, arterial blood pressure, oxygen saturation, and laboratory values were observed following administration of SH L 643A. Neither acute nor late-phase adverse effects were observed in this study.

10.6 Liposomes

10.6.1 Background

Liposomes are vesicles of varying size consisting of a spherical lipid bilayer and an aqueous inner compartment. Most of the lipids used for biomedical applications are natural, biodegradable, nontoxic, and nonimmunogenic [106]. Because lipids are amphiphilic, composed of a hydrophilic head group and a hydrophobic tail, lipid nanoparticles can act as nanocarriers for both hydrophilic and hydrophobic compounds. Since Alec Bangham's discovery more than 40 years ago, liposomes have become a pharmaceutical carrier of choice [107]. Today, a number of liposomal drugs have been approved for clinical use, with many more in clinical development.

10.6.2 Formulation

Liposomes are formed by bilayers of phospholipid molecules in an aqueous environment. They were first described by British Haematologist Sir Alec D Bangham in 1961 [108]. In the research setting, liposomes are generally formulated by sonicating lipids in an aqueous environment. In the industrial setting and for biomedical applications, liposomes are formulated using a number of techniques including the dehydration–rehydration method and extrusion technique [109].

10.6.3 Applications

Liposomes have become an established carrier and delivery system for pharmaceuticals. Liposomal formulations have been used for the treatment of cancer, infectious diseases, and autoimmune diseases [110]. Liposomes are excellent drug delivery vehicles because of their biodegradable and nontoxic nature. In addition, they can be formulated to encapsulate both hydrophilic and hydrophobic compounds. Several liposome formulations of chemotherapy have been approved by the FDA or have entered preclinical/clinical evaluations. The pegylated liposomal formulation of doxorubicin, Doxil, has been approved for the treatment of ovarian cancer. Daunoxome, a liposomal formulation of daunorubicin, has also been approved for the treatment of Kaposi's sarcoma. In addition to chemotherapeutics, liposomes have also been used to deliver photo-dynamic therapy (PDT) agents. FDA has approved the liposomal formulation of benzoporphyrin derivative monoacid ring A (Visudyne; Novartis) for the treatment of macular degeneration. This drug was also shown to be effective against tumors in sarcoma-bearing mice [111]. Igarashi et al. showed that liposomal formulation enhances the photofrin efficacy using a murine xenograft model of gastric cancer [112].

Since liposomes can encapsulate hydrophilic molecules, they have been utilized for the delivery of nucleic acid therapeutics. Matsuura et al. used cationic liposomes for gene delivery [113]. Brignole et al. delivered antisense oligonucleotides to neuroblastoma cells in vitro and in vivo using liposomes [114]. Liposomal delivery of siRNA has also shown promising results [115]. A complete list of liposomal drugs undergoing clinical investigation for cancer treatment is listed in Table 10.1.

Liposomes have also been utilized for the delivery of imaging agents. Weissig et al. generated gadolinium-loaded liposomes as MRI contrast agents of the blood pool [116]. Liposomes loaded with radionuclides, such as ^{99}Tc and ^{111}In, have been used for scintigraphy tumor imaging [117]. CT contrast agents have also been formulated using liposomes and PEGylated liposomes; they have favorable biodistribution and imaging potential in animal studies [118].

Table 10.1 Liposomal cancer therapies that are approved or undergoing clinical investigation

Composition	Trade name	Indications	Admin	Status	Company
Liposomal daunorubicin	DaunoXome	Kaposi's sarcoma	IV	Approved	Gilead
Liposomal cytarabine	DepoCyt	Malignant lymphomatous Meningitis	IT	Approved	SkyePharma
Liposomal doxorubicin	Myocet Sarcodox-ome	Metastatic breast cancer Soft tissue sarcoma	IV	Myocet is approved for breast cancer treatment. Sarcodoxome-Phase I/II	Zeneus (Myocet) GP-Pharm (Sarcodoxome)
Liposomal cisplatin	SLIT Cisplatin	Metastatic osteogenic sarcoma to the lung	Aerosol	Phase II	Transave
Liposomal vincristine	OncoTCS	Non-Hodgkins Lymphoma	IV	Phase II/III	Inex, Enzon
Liposomal complex with DNA plasmid encoding HLA-B7 and β2 globulin	Allovectin-7	Local and metastatic cancers	IL	Phase III	Vical
Liposomal cis-bis-neodocaneoato-trans-R,R-1,2-diaminocyclohexane platinum (II)	Aroplatin	Advanced colorectal cancer	IV	Phase II	Antigenics
Liposomal lurtotecan	OSI-211	Ovarian cancer	IV	Phase II	OSI Pharmaceuticals
Liposomal annamycin	L-Annamycin	Acute lymphocytic leukemia Acute myeloid leukemia	IV	Phase I	Callisto
PEG-liposomal doxorubicin	Doxil/Caelyx	Kaposi's sarcoma, metastatic breast cancer, metastatic ovarian cancer	IV	Approved	Ortho Biotech, Schering-Plough

10.6.4 Toxicity

The major components of the liposomes are phospholipids, which are naturally occurring, biocompatible and biodegradable. In vitro and in vivo testing of these phospholipids have revealed them to be safe for clinical use [119,120]. However, additional molecules used to modify the liposomes for longer circulation and stability may increase the toxic potential of liposomes [121]. Toffano et al. showed that intravenously injected liposomes containing phosphatidylserine can depress brain energy metabolism and interfere with brain catecholamine metabolism [122]. The toxicity has been attributed to the formation of lysophosphatidylserine [123]. Zbinden et al. showed that inclusion of negatively charged phosphatidic acid in liposomes accelerated the clotting cascade in guinea pigs [124]. Adams et al. demonstrated that while liposomes with phosphatidic acid and dipalmitoyl lecithin had minimal effects when injected intracerebrally into mice, liposomes with dicetylphosphate or stearylamine lead to epileptic seizures and some deaths [125].

Incorporating saturated phospholipids, cholesterol, or sphingomyelin into liposomes can increase liposome stability and alter the pharmacokinetics of the drug. In the case of antibiotic amphotericin B, saturated phospholipid liposomes had a much better toxicity profile when compared to the free drug or unsaturated liposomal formulation [126]. However, stable liposomes taken up by the phagocytes can lead to phospholipids overloading and thus inhibit phagocyte function in a dose dependent manner [127]. Allen et al. demonstrated that the impairment of phagocyte function is dependent on the liposome size and composition, dose and the presence of lipid peroxides [128]. Sphingomyelin-containing liposomes produced the greatest reticular endothelial blockade, distearoylphosphatidylcholine-cholesterol liposomes were intermediate and egg phosphatidylcholine-cholesterol liposomes produced the least impairment in mice. Allen and Smuckler also showed that repeated injection of liposomes can result in hepatic granuloma and liver inflammation in mice [129]. Weereratne et al. demonstrated that repeated injection of sphingomyelin containing liposomes can lead to hepatosplenomegaly in mice [130].

To assess the toxicity of liposomes, Parnham and Wetzig described in vitro and in vivo tests that are necessary based on prior toxicity studies [119]. The tests include erythrocyte hemolysis, platelet aggregation, cytotoxicity to mouse leukemia cells (L1210), phagocytic index, pyrogenicity, acute parenteral toxicity, systemic and local tolerability. The erythrocyte hemolysis test was first described by Yoshihara et al., who studied the hemolytic activity of stearylamine-containing liposomes. They showed that the hemolytic activity of stearylamine-liposome was markedly influenced by the composition of hydrocarbon chains of the phospholipids and that the activity can be inhibited by addition of liposomes containing negatively charged phospholipids. Several studies showed that liposomes can affect the growth and viability of L1210 leukemic cells [131]. This resulted in the use of L1210 cells for cytotoxicity assessment.

Today, there are many liposomal formulations of drugs that are approved by the FDA, with many more in clinical development. Early success of liposomal drugs such as liposomal amphoterin B and liposomal doxorubicin lead to continued high interest in liposome development. Although we cannot review all the toxicity data on liposomal cancer therapies, we will review the three formulations that have been approved by the FDA here.

Doxil/Caelyx is a PEGylated liposomal formulation of doxorubicin (Dox) with nanoparticle size around 100 nm. In two Phase I studies with 56 patients, the toxicity profile of liposomal Dox differed prominently from that of the free drug administered by bolus or rapid infusion and resembles that of continuous infusion [132]. The two dose-limiting side effects were hand–foot (H–F) syndrome and stomatitis. In an analysis of three open-label Phase II studies with 219 patients, the common adverse effects were palmar-plantar erythrodysesthesia (PPE), asthenia, stomatitis, and leucopenia/neutropenia [133]. Most importantly, the incidence of cardiac toxicity was markedly less than expected at comparable doses of free Dox. Phase III trials confirmed the prior findings, especially the lower cardiac toxicity of liposomal Dox [134].

DaunoXome is a liposomal formulation of daunorubicin which has been approved for the treatment of Kaposi's sarcoma. In several Phase I and Phase II studies, DaunoXome showed less cardiac toxicity when compared to free daunorubicin [135,136]. There was also no significant alopecia, mucositis, or vomiting. The common side effects with DaunoXome were neutropenia, and mild to moderate fatigue, nausea, and diarrhea [135,137]. In a randomized Phase III trial of liposomal daunorubicin vs. doxorubicin, bleomycin, and vincristine in 232 patients, liposomal daunorubicin cause less alopecia and neuropathy but more neutropenia [138].

DepoCyt is a liposomal formulation of cytarabine, which has been approved for the treatment of lymphomatous meningitis. In an open-label, parallel group multi-center trial with 28 patients comparing DepoCyt and free cytarabine, DepoCyt had better tolerability and there was no significant difference in toxicity profile between the drugs [139]. In another open-label trial of DepoCyt with 110 patients, the main side effects for DepoCyt were diverse events such as headache and arachnoiditis, which were low grade, transient, and reversible [140]. No cumulative toxicity was observed.

10.7 Polymeric Nanoparticles

10.7.1 Background

Polymeric nanoparticles are promising nanocarriers for therapeutics delivery, with a number of formulations in clinical use and many more in clinical trials [141,142]. Polymeric nanoparticles have several advantages, including controlled drug release, tissue penetrating ability, and reduced toxicity. Most importantly, their size, stability, loading capacity, and release kinetics of the drugs can be modulated by the structures of the particles. Furthermore, polymeric nanoparticles can be engineered to be biologically stealth and target specific tissue.

10.7.2 Synthesis

Polymeric nanoparticles include polymer-drug conjugates and polymeric micelles. Polymer-therapeutics conjugates are generally formed through direct chemical conjugation between the therapeutic agent and a polymer. Polymers have been conjugated to protein drugs, small molecules drugs, and oligonucleotides therapeutic agents [143]. Polymeric micelles are routinely synthesized by emulsion and precipitation methods, depending on the hydrophilicity nature of the therapeutic agents to be encapsulated [142]. Hydrophilic therapeutic agents are typically encapsulated into polymeric nanoparticles by water-in-oil-in-water double emulsion method [144]. In this approach, hydrophilic agents dissolved in aqueous phase are first emulsified in a non-polar organic solvent containing the biopolymer. The aqueous nanodroplets in organic solvent are then emulsified in water and form polymeric nanoparticles with an aqueous core. Hydrophobic compounds can be encapsulated into nanoparticles by nanoprecipitation [145]. In this approach, both the therapeutic agents and polymers are dissolved in an organic solvent that is miscible with water. Polymeric nanoparticles are formed by self-assembly of amphiphilic block-copolymers consisting of two or more hydrophilic and hydrophobic polymer chains. The hydrophobic compartment forms the nanoparticle core and can be used to encapsulate therapeutic agents. Poly(D,L-lactic acid), poly(D,L-glycolic acid), poly(ε-caprolactone), and their copolymers at various molar ratios either diblocked or multiblocked with poly(ethylene glycol) are the most commonly used biodegradable polymers to form polymeric nanoparticles [142,146].

10.7.3 Applications

Polymeric nanoparticles (NPs) have been used extensively for delivery of therapeutics to cancer tissue. Today, there are two polymer-drug conjugates and one polymeric micelle therapeutic approved for cancer treatment. In addition, there are many more in clinical development. Table 10.2 lists the polymeric nanoparticles approved for clinical use or in clinical trials.

PEG-L-asparaginase (Oncaspar) has been approved for the treatment of acute lymphoblastic leukemia. PEG-GCSF (Neublasta) has been approved for the treatment of neutropenia associated with chemotherapy. Both have demonstrated superior efficacy and toxicity profiles when compared to their non-conjugate counterpart. Recently, Methoxy-PEG-poly(D,L-lactide) paclitaxel (Genexol-PM), became the first clinical approved polymeric micelle. It was approved for the treatment of metastatic breast cancer.

10.7.4 Toxicity

Over the past several decades, biodegradable and biocompatible polymers have been utilized for drug delivery. A large amount of literature is available on protein and drug pegylation [147]. Polymers such as PEG, Poly(D,L-lactic acid),

Table 10.2 Polymeric nanoparticles for cancer therapy approved for clinical use or in clinical trials

Composition	Trade name	Indications	Admin	Status	Company
PEG-L-asparaginase	Oncaspar	Acute lymphoblastic leukemia	IV, IM	Approved	Enzon
PEG-GCSF	Neulasta	Neutropenia associated with cancer therapy	SC	Approved	Amgen
HPMA copolymer-DACH-platinate	ProLindac	Ovarian cancers	IV	Phase II	Access Pharmaceuticals
Methoxy-PEG-poly(D,L-lactide) paclitaxel	Genexol-PM	Metastatic breast cancer	IV	Approve in Korea	Samyang Corp
PEG-arginine deaminase	Hepacid	Hepatocellular carcinoma	IV	Phase I/II	Phoenix
PEG-camptothecin	Prothecan	Various cancers	IV	Phase I/II	Enzon
Pluronic block-copolymer doxorubicin	SP1049C	Oesophageal carcinoma	IV	Phase II	Supratek Pharma
Polycyclodextrin camptothecin	IT-101	Metastatic solid tumors	IV	Phase I	Insert Therapeutics
Polyglutamate camptothecin	CT-2106	Colorectal and ovarian cancers	IV	Phase I/II	Cell Therapeutics
Polyglutamate paclitaxel	Xyotax	Non-small cell lung cancer, ovarian cancer	IV	Phase III	Cell Therapeutics
Poly(iso-hexyl-cyanoacrylate) doxorubicin	Transdrug	Hepatocellular carcinoma	IA	Phase I/II	BioAlliance Pharma

poly(D,L-glycolic acid), and poly(ε-caprolactone) have been approved for clinical use in their macro-formulations [148]. However, there is considerably less safety data on polymeric nanoformulations.

Polyethylene glycol (PEG) is one of the most commonly used polymers for nanoparticle drug delivery. It is made up of identical ethylene glycol subunits. In their nonconjugated form, PEGs are widely used as excipients for medicines, and are also regularly used in nonpharmaceutical products including toothpaste, shampoo, food, drinks, and deodorants [149]. PEG has been shown to be non-toxic except in very high concentrations (600 mM) which lead to renal toxicity [150]. Abuchowski et al. were the first to show that PEGylation of proteins reduces the proteins' immunogenicity [151]. PEGylation of drugs can enhance the drug's stability and solubility in the plasma, while lowering drug toxicity [147,152]. PEGylated proteins were the first type of PEG conjugates evaluated for therapeutic applications.

Clinical toxicology studies performed with PEGylated proteins have not revealed any PEG-specific toxicities. For example, in clinical trials Peg-Intron (PEGylated interferon α_{2a}) did not show unique toxicity due to PEGylation [153].

PEG-L-asparaginase (Oncaspar) is the first polymeric formulation approved (1994) by the FDA for cancer therapy. L-Asparaginase is an effective antineoplastic agent for acute lymphoblastic leukemia [154]. However, the clinical use of unmodified L-asparaginase is limited by the high rate (approximately 25%) of allergic reactions [155]. Modification of L-asparaginase both lowered the hypersensitivity reaction rate and increased the drug's plasma half-life. In a Phase I trial, PEG-L-asparaginase showed a plasma half-life of 357 h compared to 20 h for unmodified L-asparaginase [156]. Furthermore, only 3 out of the 27 patients developed serious allergic reactions. In a multicenter Phase II open label clinical trial (ASP-201A), 21 patients received a single dose of PEG-L-asparaginase [157]. There was no anaphylactic reaction and the incidence of hyperglycemia and pancreatitis were lower than historical data for L-asparaginase. Overall, PEG-L-asparaginase has lower incidence of pancreatitis, hyperglycemia, neurological dysfunction, and hypersensitivity reactions, but higher incidence of hepatitis [154].

Another PEG conjugate, PEG-GCSF, has been approved for prevention of neutropenia associated with cancer therapy [158]. PEGylation of GCSF has reduced the frequency of administration without incurring additional toxicity [159]. Two other PEG conjugates, PEG-IFNα 2a and PEG-IFNα 2b, have been approved for the treatment of hepatitis B and C. They are also being evaluated in Phase I/II clinical trials for melanoma, chronic myeloid leukemia, renal cell carcinoma, and multiple myeloma.

N-(2-hydroxypropyl)methacrylamide (HPMA) is another polymer that has been used to formulate polymeric-drug conjugates. HPMA, like PEG, can lower the immunogenicity of the drug [160]. HPMA conjugates also demonstrate EPR-mediated tumor targeting and improve the antitumor activity of the drug [161]. Phase I evaluation of HPMA copolymer-Gly-Phe-Leu-Gly-doxorubicin (PK1; FCE28068) showed dose-limiting toxicities including neutropenia and mucositits, typical for anthracyclines. However, there was no cardiotoxicity observed despite cumulative dose up to 1680 mg per m^2 [162]. Another HPMA doxorubicin conjugate, HPMA-Gly-Phe-Leu-Gly-doxorubicin-galactosamine (PK2; FCE28069) has also been evaluated for hepatocellular carcinoma. The galactosamine moiety was designed to target the hepatocyte asialoglycoprotein receptor (ASGR) [163]. Although it showed more toxicity than PK1 in a Phase I/II clinical trial, patients had partial responses and stable disease after the treatment [163]. Phase I evaluations of two HPMA conjugates, one containing paclitaxel (PNU166945) and the other containing camptothecin (MAG-CPT), were disappointing. Neither showed improved toxicity profiles nor improved efficacy [164]. Two other HPMA platinate conjugates, AP5280 containing a malonate ligand and AP5346 containing a DACH platinate, had better success. Both showed reduced platinum-related toxicity [165]. In addition to PEG and HPMA conjugates, polymers such as polyglutamate and dextran have been used to formulate polymer-drug nanoparticles.

Genexol-PM is a polymeric micelle formulation of paclitaxel. It was recently approved for the treatment of metastatic breast cancer in Korea. In a Phase I clinical trial involving 21 patients, it showed favorable toxicity profile when compared to palitaxel alone [166]. There was no hypersensitivity reaction observed during the trial. The most common side effects were neuropathy and myelgia. Four patients developed grade 3–4 side effects, which included neutropenia, polyneuropathy, and myelgia. In as Phase II trial with 41 women with metastatic breast cancer, Genexol showed an overall response rate of 58.5% with 5 complete responses and 19 partial responses [167]. Among all, 51.2% patients developed grade 3 sensory peripheral neuropathy, 8 patients (19.5%) experienced hypersensitivity reactions and 68.3% of the patients experienced grade 3 or 4 neutropenia.

There are many polymeric micelles are in preclinical development. Polyesters such as poly (D,L-lactide) (PLA) and poly(glycolide) (PGA) and their copolymers (PLGA) represent the class of materials most commonly used for these nanoparticles. The toxicity of these polymers alone has been well studied [168]. As more polymeric micelles enter clinical evaluation, we will better understand the toxicities of this nanoparticles platform.

10.8 Albumin-bound Paclitaxel (Abraxane)

Albumin-bound Paclitaxel (Abraxane) is a nanoparticle formulation of paclitaxel that was approved for the treatment of metastatic breast cancer. It is distinct as it does not fall into the previously mentioned nanoparticle platforms. Instead of using Cremophor-EL, Abraxane is formulated using human serum albumin and is prepared by high pressure homogenization [169]. The preparation results in nanosized particles of albumin-bound paclitaxel.

Since Abraxane is Cremophor free, it lowers the hypersensitivity reaction generally associated with Cremophor. In a Phase I trial with 19 patients, no acute hypersensitivity reaction was observed [169]. Dose-limiting toxicity consisted of sensory neuropathy, stomatitis, and superficial keratopathy. In a multi-center Phase II clinical trial with 63 patients, Abraxane achieved an overall response rate of 48% [170]. No severe hypersensitivity reactions were observed despite the lack of premedication. Toxicities observed were typical of paclitaxel and included grade 4 neutropenia (24%), grade 3 sensory neuropathy (11%), and grade 4 febrile neutropenia (5%). In a phase III trial with 454 patients, Abraxane demonstrated higher response rates when compared to Cremophor formulated paclitaxel (Taxol) (33% vs. 19%) [171]. The incidence of neutropenia for Abraxane was 9% where that of Taxol was 22%, despite a 49% higher paclitaxel dose for Abraxane. There was no hypersensitivity reaction in the Abraxane arm.

10.9 Summary

Nanoparticles hold great potential as cancer diagnostics and therapeutics, as evidenced by the large number of nanotherapeutics in preclinical and clinical development today. While some nanoparticle systems such as liposomes and polymer conjugates have well-established toxicity profiles, other nanoparticle platforms – including dendrimers, QDs, and polymeric micelles need further investigation. There is a pressing need for systematic evaluations of the toxicity profiles of these nanoparticles, since existing toxicity data are incomplete and have been conducted under different experimental conditions. Each new nanoparticle should be evaluated for in vitro and in vivo cell toxicity, in vivo biodistribution, immunogenicity, and long-term in vivo toxicity. Most importantly, these experiments should be conducted under standard experimental conditions, which may be formulated via consensus by the research community. As we understand more about the toxicities of these nanoparticle systems, we will be able to fully realize their clinical potential. We are optimistic that with further research and development, nanoparticle therapeutic and diagnostic technologies will have an enormous impact on medicine.

Acknowledgments We thank Drs. Philip Kantoff and Neil Bander for helpful discussions throughout this study. This work was supported by National Institutes of Health Grants CA119349 and EB003647 and Koch–Prostate Cancer Foundation Award in Nanotherapeutics.

References

[1] American Cancer Society (2007) Atalanta: American Cancer Society: 2007.
[2] Moses C, Kent E, Boatman JB (1955) Cancer 8(2):417.
[3] Zimmer A (1999) Methods 18(3):286.
[4] Yuan F, Leunig M, Huang SK et al (1994) Cancer Res 54(13):3352; Yuan F, Leunig M, Berk DA et al (1993), Microvasc Res 45(3):269.
[5] Gilchrist RK, Medal R, Shorey WD et al (1957) Ann Surg 146(4):596.
[6] Babes L, Denizot B, Tanguy G et al (1999) J Colloid Interface Sci 212(2):474.
[7] Gillis P, Roch A, Brooks RA (1999) J Magn Reson 137(2):402; Muller RN, Gillis P, Moiny F et al (1991) Magn Reson Med 22(2):178.
[8] Gupta AK, Gupta M (2005) Biomaterials 26(18):3995.
[9] Benderbous S, Corot C, Jacobs P et al (2003) Acad Radiol 3 Suppl 2, S292.
[10] Harisinghani MG, Barentsz J, Hahn PF et al (2003) N Engl J Med 348(25):2491.
[11] Weissleder R, Bogdanov A, Papisov M (1996) Magn Reson Q 8(1):55.
[12] Mitsumori M, Hiraoka M, Shibata T et al (1996) Hepatogastroenterology 43(12):1431.
[13] Kohler N, Sun C, Fichtenholtz A et al (2006) Small 2(6):785.
[14] Wust P, Gneveckow U, Johannsen M et al (2006) Int J Hyperthermia 22(8):673.
[15] Wisner ER, Amparo EG, Vera DR et al (1995) J Comput Assist Tomogr 19(2):211; Chouly C, Pouliquen D, Lucet I et al (1996) J Microencapsul 13(3):245; Rety F, Clement O, Siauve N et al (2000) J Magn Reson Imaging 12(5):734; Allkemper T, Bremer C, Matuszewski L et al (2002) Radiology 223(2):432.
[16] McLachlan SJ, Morris MR, Lucas MA et al (1994) J Magn Reson Imaging 4(3):301.
[17] Taupitz M, Wagner S, Schnorr J et al (2004) Invest Radiol 39(7):394.
[18] Li W, Tutton S, Vu AT et al (2005) J Magn Reson Imaging 21(1):46.
[19] Pouliquen D, Le Jeune JJ, Perdrisot R et al (1991) Magn Reson Imaging 9(3):275.
[20] Schulze E, Ferrucci JT, Jr, Poss K et al (1995) Invest Radiol 30(10):604.
[21] Clement O, Siauve N, Cuenod CA et al (1999) J Comput Assist Tomogr 23 Suppl 1, S45.
[22] Briley-Saebo K, Bjornerud A, Grant D et al (2004) Cell Tissue Res 316(3):315.
[23] Taylor AM, Panting JR, Keegan J et al (1999) J Cardiovasc Magn Reson 1(1):23.
[24] Bonnemain B (1998) J Drug Target 6(3):167.
[25] Anastase S, Letourneur D, Jozefonvicz J (1996) J Chromatogr B Biomed Appl 686(2):141.
[26] Zinderman CE, Landow L, Wise RP (2006) J Vasc Surg 43(5):1004.

[27] Stroh A, Zimmer C, Gutzeit C et al (2004) Free Radic Biol Med 36(8):976.
[28] Raynal I, Prigent P, Peyramaure S et al (2004) Invest Radiol 39(1):56.
[29] Pisanic TR, 2nd, Blackwell JD, Shubayev VI et al (2007) Biomaterials 28(16):2572.
[30] Ros PR, Freeny PC, Harms SE et al (1995) Radiology 196(2):481.
[31] Kopp AF, Laniado M, Dammann F et al (1997) Radiology 204(3):749; Reimer P, Rummeny EJ, Daldrup HE et al (1995) Radiology 195(2):489.
[32] Sharma R, Saini S, Ros PR et al (1999) J Magn Reson Imaging 9(2):291.
[33] Johannsen M, Gneveckow U, Eckelt L et al (2005) Int J Hyperthermia 21(7):637.
[34] Maier-Hauff K, Rothe R, Scholz R et al (2007) J Neurooncol 81(1):53.
[35] Gumpel JM (1973) Ann Rheum Dis 32 Suppl 6, Suppl:29; Manadan AM, Sequeira W, Block JA (2006) Am J Ther 13(1):72; Sautner J, Leeb BF (2005) Internist (Berl) 46(12):1399; Banas M, Kucharz EJ (2005) Pol Arch Med Wewn 113(3):267.
[36] Muller JH (1963) Am J Roentgenol Radium Ther Nucl Med 89:533; Riccioni N (1969) J Nucl Biol Med 13(4):160; Volkova MA, Borisov EA, Pelman SG (1974) Radiobiol Radiother (Berl) 15(3):356; Volkova MA, Borisov EA, Pelman SG (1974) Radiobiol Radiother (Berl) 15(3):365; Rosenshein NB, Leichner PK, Vogelsang G (1979) Obstet Gynecol Surv 34(9):708.
[37] Paciotti GF, Myer L, Weinreich D et al (2004) Drug Deliv 11(3):169.
[38] Zhou HS, Honma II, Komiyama H et al (1994) Phys Rev B Condens Matter. 50(16):12052.
[39] Oldenburg SJ, Averitt RD, Westcott SL et al (1998) Chem Phys Lett 288 (2–4) 243.
[40] Sokolov K, Follen M, Aaron J et al (2003) Cancer Res 63(9):1999.
[41] Hainfeld JF, Slatkin DN, Smilowitz HM (2004) Phys Med Biol 49(18):N309.
[42] Ku G, Wang LV (2005) Opt Lett 30(5):507; Wu CF, Liang XP, Jiang HB (2005) Opt Commun 253(1–3) 214.
[43] Loo C, Lin A, Hirsch L et al (2004) Technol Cancer Res Treat 3(1):33.
[44] Hirsch LR, Stafford RJ, Bankson JA et al (2003) Proc Natl Acad Sci USA 100(23):13549.
[45] O'Neal DP, Hirsch LR, Halas NJ et al (2004) Cancer Lett 209(2):171.
[46] Merchant B, Biologicals 26(1):49 (1998).
[47] Russell MA, King LE, Jr, Boyd AS (1996) N Engl J Med 334(9):603.
[48] Antonovych TT (1981) Ann Clin Lab Sci 11(5):386.
[49] Lewis AJ, Walz DT (1982) Prog Med Chem 19:1.
[50] Smith PM, Smith EM, Gottlieb NL (1973) J Lab Clin Med 82(6):930.
[51] Gottlieb NL, Smith PM, Penneys NS et al (1974) Arthritis Rheum 17(1):56.
[52] Gottlieb NL, Smith PM, Smith EM (1972) Arthritis Rheum 15(1):16.
[53] Freyberg RH, Block WD, Levey S (1941) J Clin Invest 20(4):401.
[54] Rapson WS (1985) Contact Dermatitis 13(2):56.
[55] Brown DH, Smith WE, Fox P et al (1982) Inorg Chim Acta-Bioinorg Chem 67(1):27.
[56] Clark P, Tugwell P, Bennet K et al (2000) Cochrane Database Syst Rev (2):CD000520.
[57] Penneys NS, Ackerman AB, Gottlieb NL (1974) Arch Dermatol 109(3):372.
[58] Morley TF, Komansky HJ, Adelizzi RA et al (1984) Eur J Respir Dis 65(8):627; Kazantzis G (1978) Environ Health Perspect 25:111.
[59] Isab AA, Sadler PJ (1977) Biochim Biophys Acta 492(2):322.
[60] Schuhmann D, Kubicka-Muranyi M, Mirtschewa J et al (1990) J Immunol 145(7):2132.
[61] Connor EE, Mwamuka J, Gole A et al (2005) Small 1(3):325.
[62] Goodman CM, McCusker CD, Yilmaz T et al (2004) Bioconjugate Chem 15(4):897.
[63] Shukla R, Bansal V, Chaudhary M et al (2005) Langmuir 21(23):10644.
[64] Bruchez M, Jr., Moronne M, Gin P et al (1998) Science 281(5385):2013.
[65] Dabbousi BO, RodriguezViejo J, Mikulec FV et al (1997) J Phys Chem B 101(46):9463.
[66] Hines MA, Guyot-Sionnest P (1996) J Phys Chem 100(2):468.
[67] Kim S, Fisher B, Eisler HJ et al (2003) J Am Chem Soc 125(38):11466.
[68] Chan WCW, Nie SM (1998) Science 281(5385):2016; Pathak S, Choi SK, Arnheim N et al (2001) J Am Chem Soc 123(17):4103; Guo W, Li JJ, Wang YA et al (2003) J Am Chem Soc 125(13):3901; Pinaud F, King D, Moore HP et al (2004) J Am Chem

Soc 126(19):6115; Gao XH, Chan WCW, Nie SM (2002) J Biomedical Optics 7(4):532; Dubertret B, Skourides P, Norris DJ et al (2002) Science 298(5599):1759; Gerion D, Pinaud F, Williams SC et al (2001) J Phys Chem B 105(37):8861.

[69] Dwarakanath S, Bruno JG, Shastry A et al (2004) Biochem Biophys Res Commun 325(3):739.

[70] Kim S, Lim YT, Soltesz EG et al (2004) Nat Biotechnol 22(1):93.

[71] Gao X, Cui Y, Levenson RM et al (2004) Nat Biotechnol 22(8):969.

[72] Samia AC, Dayal S, Burda C (2006) Photochem Photobiol 82(3):617.

[73] Bakalova R, Ohba H, Zhelev Z et al (2004) Nat Biotechnol 22(11):1360.

[74] Hardman R (2006) Environ Health Perspect 114(2):165.

[75] Nath R, Prasad R, Palinal VK et al (1984) Prog in Food and Nutrition Sci 8(1–2):109.

[76] Derfus AM, Chan WCW, Bhatia SN (2004) Nano Lett 4(1):11; Kirchner C, Liedl T, Kudera S et al (2005) Nano Lett 5(2):331.

[77] Hoshino A, Fujioka K, Oku T et al (2004) Nano Lett 4(11):2163.

[78] Lovric J, Bazzi HS, Cuie Y et al (2005) J Mol Med 83(5):377.

[79] Chang E, Thekkek N, Yu WW et al (2006) Small 2(12):1412.

[80] Yu WW, Chang E, Colvin VL, Drezek R (2005) J Biomed Nanotechnol 1:397.

[81] Choi AO, Cho SJ, Desbarats J et al (2007) J Nanobiotechnol 5:1.

[82] Chan WH, Shiao NH, Lu PZ (2006) Toxicol Lett 167(3):191.

[83] Ballou B, Ernst LA, Andreko S et al (2007) Bioconjug Chem 18(2):389; Ballou B, Lagerholm BC, Ernst LA et al (2004) Bioconjug Chem 15(1):79; Akerman ME, Chan WC, Laakkonen P et al (2002) Proc Natl Acad Sci USA 99(20):12617; Larson DR, Zipfel WR, Williams RM et al (2003) Sci 300(5624):1434; Dubertret B, Skourides P, Norris DJ et al (2002) Sci 298(5599):1759; Jaiswal JK, Mattoussi H, Mauro JM et al (2003) Nat Biotechnol 21(1):47; Voura EB, Jaiswal JK, Mattoussi H et al (2004) Nat Med 10(9):993.

[84] Svenson S, Tomalia DA (2005) Adv Drug Deliv Rev 57(15):2106.

[85] Tomalia DA (1996) Macromol Symposia 101:243.

[86] Frechet JMJ, Hawker CJ (1990) J Am Chem Soc 112:7638.

[87] Bosman AW, Janssen HM, Meijer EW (1999) Chem Rev 99(7):1665.

[88] Malik N, Evagorou EG, Duncan R (1999) Anticancer Drugs 10(8):767.

[89] Wu G, Yang W, Barth RF et al (2007) Clin Cancer Res 13(4):1260.

[90] Backer MV, Gaynutdinov TI, Patel V et al (2005) Mol Cancer Ther 4(9):1423.

[91] Barth RF, Wu G, Yang W et al (2004) Appl Radiat Isot 61(5):899; Yang W, Barth RF, Wu G et al (2004) Appl Radiat Isot 61(5):981.

[92] Eichman JD, Bielinska AU, Kukowska-Latallo JF et al (2000) Pharm Sci Technolo Today 3(7):232; Bielinska AU, Chen C, Johnson J et al (1999) Bioconjug Chem 10(5):843; Hughes JA, Aronsohn AI, Avrutskaya AV et al (1996) Pharm Res 13(3):404; Juliano RL, Alahari S, Yoo H et al (1999) Pharm Res 16(4):494; Dass CR (2002) J Pharm Pharmacol 54(1):3; Juliano RL (2006) Ann N Y Acad Sci 1082:18; Kang H, DeLong R, Fisher MH et al (2005) Pharm Res 22(12):2099.

[93] Kobayashi H, Kawamoto S, Jo SK et al (2003) Bioconjug Chem 14(2):388; Roberts HC, Saeed M, Roberts TP et al (1999) J Magn Reson Imaging 9(2):204; Langereis S, de Lussanet QG, van Genderen MH et al (2006) NMR Biomed 19(1):133.

[94] Roberts JC, Bhalgat MK, Zera RT (1996) J Biomed Mater Res 30(1):53.

[95] Hong S, Bielinska AU, Mecke A et al (2004) Bioconjug Chem 15(4):774.

[96] Chen HT, Neerman MF, Parrish AR et al (2004) J Am Chem Soc 126(32):10044.

[97] Lee JH, Lim YB, Choi JS et al (2003) Bioconjug Chem 14(6):1214.

[98] Malik N, Wiwattanapatapee R, Klopsch R et al (2000) J Control Release 65(1–2):133.

[99] Plank C, Mechtler K, Szoka FC, Jr, et al (19966) Hum Gene Ther 7(12): 1437.

[100] Jevprasesphant R, Penny J, Attwood D et al (2003) Pharm Res 20(10):1543; Jevprasesphant R, Penny J, Jalal R et al (2003) Int J Pharm 252(1–2):263.

[101] Barth RF, Adams DM, Soloway AH et al (1994) Bioconjug Chem 5 (1): 58; Patri AK, Myc A, Beals J et al (2004) Bioconjug Chem 15(6):1174.

[102] Capala J, Barth RF, Bendayan M et al (1996) Bioconjug Chem 7(1):7.
[103] Kobayashi H, Kawamoto S, Jo SK et al (2002) Kidney Int 61 (6): 1980; Kobayashi H,
       Jo SK, Kawamoto S et al (2004) J Magn Reson Imaging 20 (3):512; Sato N, Kobayashi H,
       Hiraga A et al (2001) Magn Reson Med 46 (6):1169; Kobayashi H, Sato N, Kawamoto S
       et al (2001) Magn Reson Med 46 (3):457.
[104] Kobayashi H, Kawamoto S, Saga T et al (2001) Magn Reson Med 46(4):795.
[105] Herborn CU, Barkhausen J, Paetsch I et al (2003) Radiol 229(1):217.
[106] Voinea M, Simionescu M (2002) J Cell Mol Med 6(4):465.
[107] Torchilin VP (2005) Nat Rev Drug Discov 4(2):145.
[108] Bangham AD (1972) Annu Rev Biochem 41:753.
[109] Walde P, Ichikawa S (2001) Biomol Eng 18 (4):143; Olson F, Hunt CA, Szoka FC et al
       (1979) Biochim Biophys Acta 557 (1):9.
[110] Felnerova D, Viret JF, Gluck R et al (2004) Curr Opin Biotechnol 15(6):518.
[111] Ichikawa K, Takeuchi Y, Yonezawa S et al (2004) Cancer Lett 205(1):39.
[112] Igarashi A, Konno H, Tanaka T et al (2003) Toxicol Lett 145(2):133.
[113] Matsuura M, Yamazaki Y, Sugiyama M et al (2003) Biochim Biophys Acta 1612(2):136.
[114] Brignole C, Pagnan G, Marimpietri D et al (2003) Cancer Lett 197(1–2):231.
[115] Sioud M, Sorensen DR (2003) Biochem Biophys Res Commun 312(4):1220.
[116] Weissig VV, Babich J, Torchilin VV (2000) Colloids Surf B Biointerfaces 18(3–4):293.
[117] Ogihara-Umeda I, Sasaki T, Kojima S et al (1996) J Nucl Med 37 (2):326; Goins B, Klip-
       per R, Rudolph AS et al (1994) J Nucl Med 35 (9):1491; Proffitt RT, Williams LE, Presant
       CA et al (1983) J Nucl Med 24(1):45.
[118] Sachse A, Leike JU, Schneider T et al (1997) Invest Radiol 32 (1):44; Sachse A, Leike
       JU, Rossling GL et al (1993) Invest Radiol 28 (9):838; Schuhmann-Giampieri G, Leike J,
       Sachse A et al (1995) Pharm Res 12 (7):1065.
[119] Parnham MJ, Wetzig H (1993) Chem Phys Lipids 64 (1–3):263.
[120] Randolph WF (1983) Fed Regist 48:51149.
[121] Oussoren C, Storm G, Peters PAM, Barenholz Y (1993) CRC Press III, 345.
[122] Bruni A, Toffano G, Leon A et al (1976) Nature 260(5549):331; Toffano G, Bruni A (1980)
       Pharmacol Res Commun 12(9):829.
[123] Bigon E, Boarato E, Bruni A et al (1979) Br J Pharmacol 67(4):611; Bigon E, Bruni A,
       Mietto L et al (1980) Br J Pharmacol 69(1):11.
[124] Zbinden G, Wunderli-Allenspach H, Grimm L (1989) Toxicol 54(3):273.
[125] Adams DH, Joyce G, Richardson VJ et al (1977) J Neurol Sci 31(2):173.
[126] Juliano RL, Grant CW, Barber KR et al (1987) Mol Pharmacol 31(1):1.
[127] Senior JH (1987) Crit Rev Ther Drug Carrier Syst 3(2):123.
[128] Allen TM, Murray L, MacKeigan S et al (19684) J Pharmacol Exp Ther 229(1):267.
[129] Allen TM, Smuckler EA (1985) Res Commun Chem Pathol Pharmacol 50(2):281.
[130] Weereratne EA, Gregoriadis G, Crow J (1983) Br J Exp Pathol 64(6):670.
[131] Panzner EA, Jansons VK (1979) J Cancer Res Clin Oncol 95 (1):29; Layton D, Luckenbach
       GA, Andreesen R et al (1980) Eur J Cancer 16(12):1529; Campbell PI (1983) Cytobios 37
       (145):21.
[132] Uziely B, Jeffers S, Isacson R et al (1995) J Clin Oncol 13(7):1777.
[133] Johnston SR, Gore ME (2001) Eur J Cancer 37 Suppl 9, S8.
[134] Harris L, Batist G, Belt R et al (2002) Cancer 94(1):25; Muggia F, Hamilton A (2001) Eur
       J Cancer 37 Suppl 9, S15.
[135] Gill PS, Espina BM, Muggia F et al (1995) J Clin Oncol 13(4):996.
[136] Money-Kyrle JF, Bates F, Ready J et al (1993) Clin Oncol (R Coll Radiol) 5(6): 367;
       Guaglianone P, Chan K, DelaFlor-Weiss E et al (1994) Invest New Drugs 12 (2):103.
[137] Wittgen BP, Kunst PW, van der Born K et al (2007) Clin Cancer Res 13(8):2414.
[138] Gill PS, Wernz J, Scadden DT et al (1996) J Clin Oncol 14(8):2353.
[139] Glantz MJ, LaFollette S, Jaeckle KA et al (1999) J Clin Oncol 17(10):3110.
[140] Jaeckle KA, Batchelor T, O'Day SJ et al (2002) J Neurooncol 57(3):231.

[141] Farokhzad OC, Cheng JJ, Teply BA et al (2006) Proc Natl Acad Sci USA 103 (16):6315; Gu F, Karnik R, Wang A et al (2007) Nano Today 2(3):14.
[142] Gref R, Minamitake Y, Peracchia MT et al (1994) Science 263(5153):1600.
[143] Parveen S, Sahoo SK (2006) Clin Pharmacokinet 45(10):965.
[144] Laurencin CT, Langer R (1987) Clin Lab Med 7(2):301.
[145] Cheng J, Teply BA, Sherifi I et al (2007) Biomaterials 28(5):869.
[146] Farokhzad OC, Jon SY, Khadelmhosseini A et al (2004) Cancer Res 64(21):7668.
[147] Davis FF, Kazo GM, Nucci ML et al (1990) in Peptide and Protein Drug Delivery, edited by VHL Lee (Marcel Dekker, New York); Harris JM ed. (1992) Poly(Ethylene Glycol) Chemistry, Biotechnical and Biomedical Applications. (Plenum Press, New York); Harris JM, Zalipsky S eds. (1997) Poly(Ethylene Glycol) Chemistry and Biological Applications. (American Chemical Society, Washingston, DC).
[148] Schugens C, Grandfils C, Jerome R et al (1995) J Biomed Mater Res 29(11):1349; Ignatius AA, Claes LE (1996) Biomaterials 17 (8):831; Cordewene FW, van Geffen MF, Joziasse CA et al (2000) Biomaterials 21(23):2433.
[149] Webster R, Didier E, Harris P et al (2007) Drug Metab Dispos 35(1):9.
[150] Fruijtier-Polloth C (2005) Toxicology 214(1–2):1.
[151] Abuchowski A, McCoy JR, Palczuk NC et al (1977) J Biol Chem 252(11): 3582; Abuchowski A, van Es T, Palczuk NC et al (1977) J Biol Chem 252(11):3578; Abuchowski A, Davis FF (1979) Biochim Biophys Acta 578 (1):41.
[152] Delgado C, Francis GE, Fisher D (1992) Crit Rev Ther Drug Carrier Syst 9(3–4):249.
[153] Baker DE (2001) Rev Gastroenterol Disord 1(2):87.
[154] Narta UK, Kanwar SS, Azmi W (2007) Crit Rev Oncol Hematol 61(3):208.
[155] Hand Schumacher RE, Uren JR (1977) Cancer: A comprehensive treatise. (Plenum Press, New York).
[156] Ho DH, Brown NS, Yen A et al (1986) Drug Metab Dispos 14(3):349.
[157] Ettinger LJ, Kurtzberg J, Voute PA et al (1995) Cancer 75(5):1176.
[158] Molineux G (2004) Curr Pharm Des 10(11):1235.
[159] Johnston F, Crawford J, Blackwell S et al (2000) J Clin Oncol 18(13):2522; Green MD, Koelbl H, Baselga J et al (2003) Ann Oncol 14(1):29.
[160] Rihova B, Ulbrich K, Kopecek J et al (1983) Folia Microbiol (Praha) 28(3):217.
[161] Seymour LW, Ulbrich K, Steyger PS et al (1994) Br J Cancer 70(4):636; Duncan R, Coatsworth JK, Burtles S (1998) Hum Exp Toxicol 17 (2):93.
[162] Vasey PA, Kaye SB, Morrison R et al (1999) Clin Cancer Res 5(1):83.
[163] Seymour LW, Ferry DR, Anderson D et al (2002) J Clin Oncol 20(6):1668.
[164] Meerum Terwogt JM, ten Bokkel Huinink WW, Schellens JH et al (2001) Anticancer Drugs 12(4):315; Bissett D, Cassidy J, de Bono JS et al (2004) Br J Cancer 91(1):50; Wachters FM, Groen HJ, Maring JG et al (2004) Br J Cancer 90(12):2261; Schoemaker NE, van Kesteren C, Rosing H et al (2002) Br J Cancer 87(6):608.
[165] Gianasi E, Buckley RG, Latigo J et al (2002) J Drug Target 10(7):549; Rademaker-Lakhai JM, Terret C, Howell SB et al (2004) Clin Cancer Res 10(10):3386; Campone M, Rademaker-Lakhai JM, Bennouna J et al (2007) Cancer Chemother Pharmacol.
[166] Kim TY, Kim DW, Chung JY et al (2004) Clin Cancer Res 10(11):3708.
[167] Lee KS, Chung HC, Im SA et al (2007) Breast Cancer Res Treat.
[168] Athanasiou KA, Niederauer GG, Agrawal CM (1996) Biomaterials 17(2):93.
[169] Ibrahim NK, Desai N, Legha S et al (2002) Clin Cancer Res 8(5):1038.
[170] Ibrahim NK, Samuels B, Page R et al (2005) J Clin Oncol 23(25):6019.
[171] Gradishar WJ, Tjulandin S, Davidson N et al (2005) J Clin Oncol 23(31):7794.

Index

In the below index '*tab*' indicates entries in the table and '*fig*' indicates entries in the figure